Plato's Camera

Plato's Camera

How the Physical Brain Captures a Landscape of Abstract Universals

Paul M. Churchland

The MIT Press
Cambridge, Massachusetts
London, England

For information about special quantity discounts, please email special_sales@ mitpress.mit.edu

This book was set in Stone Sans and Stone Serif by Toppan Best-set Premedia Limited. Printed and bound in the United States of America.

Library of Congress Cataloging-in-Publication Data

Churchland, Paul M., 1942–.
Plato's camera : how the physical brain captures a landscape of abstract universals / Paul M. Churchland.
   p. cm.
Includes bibliographical references (p.   ) and index.
ISBN 978-0-262-01686-5 (hardcover : alk. paper)
1. Philosophy of mind. 2. Universals (Philosophy). 3. Cognitive neuroscience—Philosophy. I. Title.
B945.C473P63 2012
128'.2—dc23
                                                        2011021046

10 9 8 7 6 5 4 3 2 1

Contents

Preface   vii

1   Introduction: A Fast Overview   1
1   Some Parallels and Contrasts with Kant   1
2   Representations in the Brain: Ephemeral versus Enduring   4
3   Individual Learning: Slow and Structural   11
4   Individual Learning: Fast and Dynamical   16
5   Collective Learning and Cultural Transmission   25
6   Knowledge: Is It True, Justified Belief?   30

2   First-Level Learning, Part 1: Structural Changes in the Brain and the Development of Lasting Conceptual Frameworks   35
1   The Basic Organization of the Information-Processing Brain   35
2   Some Lessons from Artificial Neural Networks   38
3   Motor Coordination   45
4   More on Colors: Constancy and Compression   50
5   More on Faces: Vector Completion, Abduction, and the Capacity for 'Globally Sensitive Inference'   62
6   Neurosemantics: How the Brain Represents the World   74
7   How the Brain Does *Not* Represent: First-Order Resemblance   78
8   How the Brain Does *Not* Represent: Indicator Semantics   90
9   On the Identity/Similarity of Conceptual Frameworks across Distinct Individuals   104

3   First-Level Learning, Part 2: On the Evaluation of Maps and Their Generation by Hebbian Learning   123
1   On the Evaluation of Conceptual Frameworks: A First Pass   123
2   The Neuronal Representation of Structures Unfolding in Time   139
3   Concept Formation via Hebbian Learning: Spatial Structures   157
4   Concept Formation via Hebbian Learning: The Special Case of Temporal Structures   165
5   A Slightly More Realistic Case   170

6   In Search of Still Greater Realism   174
7   Ascending from Several Egocentric Spaces to One Allocentric Space   180

4   Second-Level Learning: Dynamical Changes in the Brain and Domain-Shifted Redeployments of Existing Concepts   187
1   The Achievement of Explanatory Understanding   187
2   On the Evaluation of Conceptual Frameworks: A Second Pass (Conceptual Redeployments)   196
3   On the Evaluation of Conceptual Frameworks: A Third Pass (Intertheoretic Reductions)   204
4   Scientific Realism and the Underdetermination of Theory by Evidence   215
5   Underdetermination Reconceived   223

5   Third-Level Learning: The Regulation and Amplification of First- and Second-Level Learning through a Growing Network of Cultural Institutions   251
1   The Role of Language in the Business of Human Cognition   251
2   The Emergence and Significance of Regulatory Mechanisms   255
3   Some Prior Takes on This Epicerebral Process   261
4   How Social-Level Institutions Steer Second-Level Learning   268
5   Situated Cognition and Cognitive Theory   274

Appendix   279
References   281
Index   287

Preface

That the *eye* is a camera is now a commonplace. It most surely is a camera, and we understand in great detail how it works. That the *brain* might be a camera is much less obvious. Indeed, the suggestion is likely to furrow one's forehead in sheer incomprehension. What could be the point of such a metaphor?

Its point is very different from the case of the eye, but the metaphor is no less apt. The eye constructs a representation, or 'takes a picture,' of the landscape or configuration of the objective *spatiotemporal particulars* currently displayed before its lens. This picture-taking process is completed in milliseconds, and the eye does it again and again, because its target landscape is typically in constant flux. The learning brain, by contrast, very slowly constructs a representation, or 'takes a picture,' of the landscape or configuration of the *abstract universals*, the *temporal invariants*, and the *enduring symmetries* that structure the objective universe of its experience. That process takes months, years, decades, and more, because these background features take time to reveal themselves in full. Moreover, the brain of each creature typically undergoes this 'picture taking' process only once, so as to produce the enduring background conceptual framework with which it will interpret its sensory experience for the rest of its life. And yet the brain manages to pull that abstract information—about the universe's timeless categorical and dynamical structure—from its extended sensory encounters with that universe, no less than the eye manages to pull a representation of the fleeting here-and-now from its current luminous input. For this reason, it is appropriate to think of the biological brain as *Plato's Camera*. This marvelous organ has the power to gain a lasting grip on those much more fundamental dimensions of reality, those dimensions that are timeless, changeless, and still.

But what manner of 'pictures' might these decidedly more abstract representations be? On this score, think *maps*. Not the two-dimensional maps that grace your automobile's glove compartment, but *high-dimensional* maps—maps with three, or a hundred, or even a million distinct dimensions, maps with extraordinary resolution and structural detail. Such maps reside—hundreds and perhaps even thousands of distinct maps—inside the brains of animals in general, and inside the brains of humans in particular. They are maps not of any geographical realities; their high-dimensionality takes them out of that comparatively meager realm. Rather, they are maps of abstract *feature domains*. They are maps of families of complex *universals* and the often intricate similarity-and-difference relations that unite and divide them. They are maps, that is, of the timeless and invariant *background structure* of the ever-changing, ever-unfolding physical universe in which all brains are constrained to operate. They are maps that constitute the 'conceptual frameworks' so familiar to us from the philosophical tradition, and so vital to any creature's comprehension of the world in which it lives.

However, and contrary to tradition, these frameworks are *not* families of predicate-like elements, united by a further family of sentence-like general commitments in which those elements appear. They are not Quinean 'webs of belief,' nor any other classical 'system-of-sentences.' Indeed, they are not linguaformal at all. Instead, these high-dimensional maplike frameworks typically consist of a large family of high-dimensional prototype points and prototype-trajectories united and mutually configured by an intricate family of similarity (i.e., proximity) and difference (i.e., distality) relations. The full range of such prototype points and trajectories represent the full range of possible *kinds of things*, and possible *kinds of processes* and *behavioral sequences*, that the creature expects, or is conceptually prepared, to encounter.

Their high-dimensionality and abstract subject-matters notwithstanding, such maps represent the world in the same fashion that any successful map represents its target domain. Specifically, there is a homomorphism or conformal relation—that is, a similarity-of-internal-structure relation—between the configuration of map-elements within the neuronal activation-space that embodies the map at issue, on the one hand, and the configuration of objective similarity-and-difference relations that structure the abstract domain of objective features or processes thereby represented, on the other. In short, the inner conceptual map 'mirrors' the outer feature-domain. Imperfectly, no doubt. But possession of the map allows the creature to anticipate at least some of the real-world features that it encounters, at least some of the time. The interactive library of such maps, possessed by any creature, constitutes its background knowledge, or better, its background understanding (for better or for worse), of the world's lasting abstract structure. The creature's subsequent pragmatic adventures depend utterly on that background understanding—on its extent and on its accuracy. Despite the urgings of some prominent Pragmatists, knowledge is not *just* a representation-free matter of being *able* or knowing *how* to behave. Our knowledge is richly representational, as we shall see, and our motor behavior depends on it profoundly. It's just that the relevant representations are not remotely propositional or linguaformal in character. Rather, they are high-dimensional geometrical manifolds. They are maps.

Maps, as we know, can be 'indexed.' That is, a point within the map can be specified, perhaps with a reaching fingertip, as the map-indexer's *current location*, within the larger range of locational possibilities comprehended by the map as a whole. The indexer's fingertip may assume a specific position on the two-dimensional highway map, a position with a unique $<x, y>$ pair of coordinates, to indicate "you are here." The abstract feature-domain maps within the brain can also be 'indexed,' this time by the activity of our *sense organs*, to indicate to the brain "you are *here*" in the space of possible objective situations. A sense organ's current activity can cause a signature *pattern*, $<x_1, x_2, \ldots, x_n>$, of n simultaneous activation-levels, across some population of n neurons that embodies the relevant map. And that activation-pattern will specify a unique position within the coordinates of that n-dimensional space, a position that represents the abstract feature currently confronted in the creature's perceptual environment. In this way, our several sense organs are continuously indexing our many feature-space maps to provide us with an unfolding understanding of our unfolding objective world-situation. Without such internal maps for our senses to index, there would be no understanding at all. We may call this the Map-Indexing Theory of Perception, and it gives a contemporary voice to Plato's further claim that a prior grasp of 'universals' is necessary for any particular case of perceptual understanding.

How those feature-domain maps are actually embodied in one's neuronal populations, thence to be indexed by subsequent sensory activity, is one of the central elements of the account of cognition to be explored in this book, as is the story of how those intricate feature-domain maps are generated or learned in the first place. This is the story of Basic-Level or First-Level Learning, to be pursued in chapters 2 and 3. (As you may imagine, Plato's own fantastical story, of a prebirth visit to a nonphysical realm, has no place in the account to be proposed.)

And there are two further kinds of learning, quite different but no less important, to be pursued in later chapters. Still, I will bring these introductory remarks to a close here. Their function has been to provoke your imagination, and to prepare your understanding, for an account of cognition that deploys an importantly new set of kinematical and dynamical resources—those of Cognitive Neuroscience—to address the questions that have long defined the discipline of Epistemology. What is knowledge? How is it acquired? How is it, or how should it be, evaluated? How does it make progress? What is its long-term destiny? By the end of this book, you will be confronting a new and unfamiliar set of answers. Whether those answers are worth embracing, I leave for you to judge. But however strange they may be, you will recognize them as brain-focused answers to long-familiar questions.

1 Introduction: A Fast Overview

1 Some Parallels and Contrasts with Kant

A novel idea is sometimes best introduced by analogy, or contrast, with ideas already familiar. Let us begin, then, with the Kantian portrait of our epistemological situation, and more specifically, with his characterization of the two faculties of empirical *intuition* and rational *judgment*. Kant argued that both of these faculties constitute a human-specific canvas on which the activities of human cognition are doomed to be painted. Space and time were claimed to be the 'pure forms' of all *sensory* intuition—the abstract background forms, that is, of all possible human sensory representations. And the organized family of the various 'pure concepts of the understanding' were claimed to provide the inevitable framework of expression for any of the *judgments* that we humans ever make about the empirical world. Accordingly, while the world-in-itself (the 'noumenal' world) is certainly not 'constructed' by us, the world-as-perceived-and-thought-by-us (the 'empirical' world of three-dimensional physical objects) does indeed display a substantial component that reflects precisely the peculiar contributions brought to the business of cognition by our own internal cognitive machinery (fig. 1.1).

Kant, of course, had an agenda that we moderns need not share: the desire to vindicate an alleged class of *synthetic a priori* truths (e.g., geometry and arithmetic), and to explain in detail how such truths are possible. Beyond this, he had commitments that we moderns may wish to deny, such as the innateness of the 'pure forms and concepts' at issue, and the implasticity of the cognitive activities that they make possible. But his portrait still constitutes a useful starting point from which a competing, and importantly different, portrait can be quickly sketched and readily grasped.

Consider, then, the possibility that human cognition involves not just *two* abstract 'spaces'—of possible human *experiences* on the one hand, and

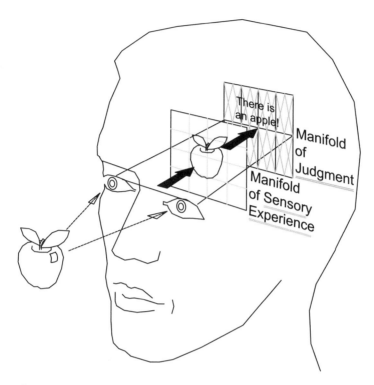

Figure 1.1
Kant's portrait of human cognition.

possible human *judgments* on the other—but rather many *hundreds*, perhaps even *thousands*, of internal cognitive 'spaces,' each of which provides a proprietary canvas on which some aspect of human cognition is continually unfolding (fig. 1.2). Consider the possibility that each such figurative cognitive space is physically embodied in the very real space of the possible *collective* activities of some proprietary population of topic-specific *neurons* within the human or animal brain.

Suppose also that the internal character of each of these representational spaces is not *fixed* by some prior decree, either divine or genetic, but is rather shaped by the extended experience of the developing animal, to reflect the peculiar empirical environment and practical needs that it encounters, and the peculiar learning procedures embodied in the brain's ongoing business of synaptic modification. These internal spaces may thus be *plastic* to varying degrees, and may hold out the promise of an enormous *range* of conceptual and perceptual possibilities for one and the same

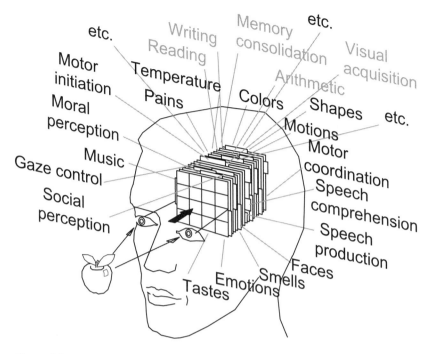

Figure 1.2
The multiple-maps portrait of human cognition.

species of creature, in stark contrast to the frozen conceptual prison contrived for us by Kant.

Consider also the possibility that the human brain devotes every bit as much of its cognitive activity to the production and administration of coherent *motor behavior* (walking, swimming, speaking, piano playing, throwing strikes, making a dinner, running a meeting) as it does to the perceptual and judgmental activities of typical concern to traditional philosophies. And note the possibility that such proprietary cognitive spaces—whose neuronal basis is located, for example, in the frontal cortex, the motor cortex, and the cerebellum—can successfully *represent* the complex motor procedures and action procedures listed above, by means of the muscle-manipulating *trajectories* of the collective neuronal activities within those spaces. Note also that such trajectories and limit-cycles within sundry *sensory* spaces can equally well represent complex *causal* processes and *periodic* phenomena, as externally encountered in perceptual experience rather than as internally generated to produce bodily behavior.

So we begin by expanding the number of representational spaces, into the hundreds and thousands, far beyond the Kantian pair. We locate them in discrete anatomical parts of the brain. We make each one of them plastic and pluripotent in its semantic content and its conceptual organization. And we reach out to include motor cognition and practical skills, along with perceptual apprehension and theoretical judgment, as equal partners in our account of human knowledge.

2   Representations in the Brain: Ephemeral versus Enduring

But we have not yet confronted the single largest contrast that we need to draw, between the Kantian portrait above and the account to be pursued in this book. For Kant, there is no question but that the fundamental unit of human cognition is the *judgment*, a unit that lives in a space of sundry logical relations with other actual and possible judgments, a unit that displays the characteristic feature of truth or falsity. However, on the account proposed in this book, there is no question but that the 'judgment'—as conceived by Kant and by centuries of other logicians—is *not* the fundamental unit of cognition, not in animals, and not in humans either. Instead, the fundamental unit of cognition—strictly, of occurrent or ephemeral cognition—is the *activation pattern* across a proprietary *population* of neurons. It is the activated *point* within any one of the many hundreds of representational *spaces* urged above.

This fundamental form of representation is one we share with all other creatures that possess a nervous system, and it does roughly the same job in every space and in every case. Such a representation lets the animal know—better, it constitutes the animal's knowledge—that its current location, in the space of possible situations comprehended by the underlying population of neurons, is *here* on the cognitive map embodied in that population. That activation point is rather like the brilliant dot of a laser pointer illuminating one tiny spot on an otherwise unlit highway map, a moving dot that continuously updates one's current position in the space of geographic possibilities represented by that map.

Since these thousands of spaces or 'maps' are all connected to one another by billions of axonal projections and trillions of synaptic junctions, such specific locational information within one map can and does provoke subsequent pointlike activations in a *sequence* of downstream representational spaces, and ultimately in one or more *motor*-representation spaces, whose unfolding activations are projected onto the body's muscle systems, thereby to generate cognitively informed behaviors.

On this view, Kantian-style 'judgments,' though entirely real, constitute an extremely peripheral form of representational activity, marginal even for adult humans, and completely absent in nonhuman animals and pre-linguistic children. How we humans manage to generate a 'language space,' to sustain our speech production and speech comprehension, is an engaging scientific question to which we shall return later in the book. For now, let me announce that, for better or for worse, the view to be explored and developed in this book is diametrically opposed to the view that humans are capable of cognition precisely because we are born with an innate 'language of thought.'

Fodor has defended this linguaformal view most trenchantly and resourcefully in recent decades, but of course the general idea goes back at least to Kant and Descartes. My own hypothesis is that all three of these acute gentlemen have been falsely taken in by what was, until recently, the only *example* of a systematic representational system available to human experience, namely, human language. Encouraged further by the structure of our own dearly beloved Folk Psychology,[1] they have wrongly read back *into* the objective phenomenon of cognition-in-general a histori-cally *accidental* structure that is idiosyncratic to a single species of animal (namely, humans), and which is of profoundly secondary importance even there. We do of course use language—a most blessed development we shall explore in due course—but language-like structures do not embody the basic machinery of cognition. Evidently they do not do so for animals, and not for humans either, because the human neuronal machinery, overall, differs from that of other animals in various small degrees, but not in fundamental kind.

An account of cognition that locates us on a continuum with all of our evolutionary brothers and sisters is thus a prime desideratum of any responsible epistemological theory. And the price we have to pay to meet it is to give up the linguaformal 'judgment' or 'proposition' as the presumed unit of knowledge or representation. But we need no longer make this sacrifice in the dark. Given the conceptual resources of modern neurobiology and cognitive neuromodeling, we are finally in a position to pursue an alternative account of cognition, one that embraces some highly specific and very different units of representation. What I hope to show in what follows is that this initial investment in an unorthodox

---

1. Folk psychology is the ordinary conceptual framework—comprehending notions like "desires that *p*," "believes that *p*," "fears that *p*," "intends that *p*"—used by lay persons to explain and predict the behavior of other humans.

assumption, concerning the nature of our primary units of representation, will yield extraordinary dividends as we proceed.

One of those dividends emerges very early in our story, for the portrait of knowledge held out to us draws a fundamental distinction between the ephemeral vehicles of our knowledge of the fleeting-here-and-now on the one hand, and the comparatively stable and enduring vehicles of our *background* knowledge of the world's-general-structure-in-space-and-time on the other. As suggested, the former vehicles are the fleeting activation patterns across a given population of neurons; they are the ever-moving, ever-jumping points of here-and-now activation within their proprietary conceptual subspaces. Think again of the moving, darting laser-dot on the otherwise unlit road map.

The latter or 'background' vehicles are entirely different. At this level of *general* knowledge, the vehicle or unit of representation is *the entire conceptual framework*. It is the *entire activation space* for the relevant population of neurons, a space that has been sculpted by months or years of learning, a space that encompasses all of the possible instances of which the creature currently has any conception. Indeed, that space is precisely the background canvas on which each fleeting *instance* of any category gets 'painted,' and that painting consists in nothing more nor less than an activation at a specific location within that sculpted space of *possible* activations. (To reprise the metaphor of the previous paragraph, the background conceptual framework is the entire *road map* at issue, the waiting space of all *possible* positions that the laser-dot might at some time illuminate.)

To illustrate with a simple and concrete example, consider the spindle-shaped space of possible *color-experiences* portrayed in figure 1.3 (plate 1). That space embodies every possible color-qualia of which the human visual system is normally capable, and that space is organized in a very specific way, a way common to all humans with normal color vision. Barring some form of color-blindness, we all share the *same* family of distance-relations and betweenness-relations that collectively locate each color-representation within that space, relative to all of its sibling color-representations. To possess this roughly double-coned space is to have the most rudimentary human knowledge of the general structure of the domain of objective colors. And to have a current activation vector at a specific point within this internal conceptual space (halfway up the central axis, for example) is to have a current representation or *experience* of a specific color (in that case, middle-*gray*). Figure 1.3 (plate 1) displays a color-coded sampling of representational positions within that space, but of course the space itself is continuous and includes positions for all of the various hues and shades

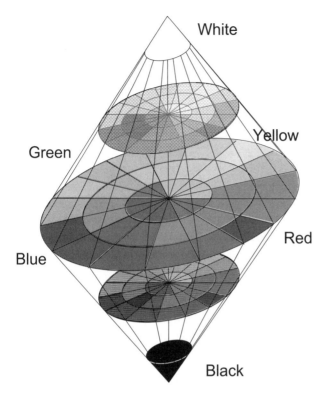

Figure 1.3
A map of the space of possible colors. See plate 1.

that occupy the various positions between the sample colors actually displayed in the diagram.

A second example concerns one's internal conceptual space for representing human *faces*, as (speculatively) portrayed in the example of figure 1.4. In fact, that three-dimensional space is a schematic of the activation space of a specific population of neurons within an *artificial* neural network, one that attempts to model the gross structure of the primary visual pathway in humans. That network was trained to discriminate faces from nonfaces, male faces from female faces, and to reidentify the faces of various named individuals across diverse photographs of each.[2]

2. See Cottrell 1991. The artificial network actually had eighty neurons at the crucial representational layer, not three, as my three-dimensional diagram would suggest. I deploy this low-dimensional fiction to help make visual sense of what is happening in the more complex case of an eighty-dimensional space.

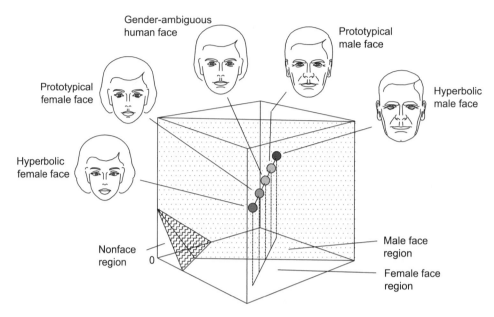

Figure 1.4
A map of the space of possible human faces.

As you can see immediately, the training process has produced a hier-archical structure of distinct representational regions within the space as a whole. Sundry nonface images presented to the network's sensory neurons get represented by sundry activation points close to the *origin* of this downstream conceptual space. By contrast, various face images, pre-sented as sensory input, get represented by various activation points within a much larger 'face region' away from the origin. That complementary region is itself split in two, into roughly equal regions for male faces and female faces, respectively. Within each gender-subspace lies a scatter of much smaller subspaces (not shown), each of which comprehends a cluster of closely proximate activation points for representing distinct sensory presentations of the face of a unique individual.

Of course, *every* point in the larger region represents a face of *some* kind or other, not just the five points illustrated. But as one moves progressively farther away from the solid straight line between the average or prototypi-cal male-face point and the average or prototypical female-face point (see again fig. 1.4), one encounters representations of faces that are increasingly nonstandard or hyperbolic in various ways. Since such outlying points

Figure 1.5
Colored afterimages due to neuronal fatigue. See plate 2.

contrast maximally with the more central points, they represent faces that 'stand out' from the crowd, faces that differ maximally from what we expect in an average, humdrum face. And just as the color space has a middle-*gray* at its central point, so our face space has a gender-*neutral* face at its central point.

The evident structure of these two spaces—for colors and for faces— reveals itself anew in some simple perceptual illusions. Fix your gaze on the × at the center of the reddish square on the right side of figure 1.5 (plate 2) and hold it for ten seconds. Then fixate on the × in the central square immediately to its left. For a few brief seconds you will perceive that hue-neutral square as being faintly green. A parallel experiment, beginning with the genuinely green square at the left of figure 1.5, will make that same central square look faintly red. If you look back to the color space of figure 1.3 (plate 1), you will notice that red and green are coded at opposite sides of that space, with maximal distance between them. Forcing oneself into a protracted period of steady neural activity, in order to code either one of these extremal colors, produces a specific pattern of short-term 'fatigue' and/or 'potentiation' across the three kinds of color-coding neurons that underlie our three-dimensional color space.

That fatigue induces those neurons, when finally relieved of their extremal burden, to fall back *past* their normal neutral points, that is, to relax back into a momentary activation pattern that briefly *mis*represents a *hue-neutral* perceptual input as if it were an input from the side of the color space exactly *opposite* from the original, fatigue-inducing stimulus. Hence the brief illusion of its exactly complementary color.

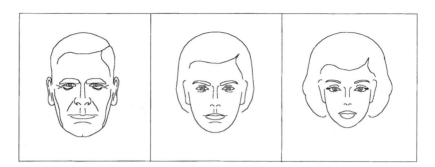

Figure 1.6
Gender-switch effect due to neuronal fatigue.

You may observe a similar short-term illusion concerning the *gender* of the face portrayed in the middle position of figure 1.6. That deliberately gender-ambiguous, vector-average, neutral human face is flanked by a hyperbolic female face on the right, and a hyperbolic male face on the left. Use your hand to cover all but the male face, and fixate on the bridge of his nose for ten seconds (do not move your eyes). Then slide your hand to the right so as to reveal (only) the neutral face, shift your fixation suddenly to the bridge of *its* nose, and make a snap judgment as to the *gender* of that (objectively neutral) face. Since hyperbolic male and female faces (like red and green colors) are coded at opposite sides of the relevant activation space, the preceding color-experiment suggests that, thanks to a comparable fatigue or saturation effect, the neutral face will now look, if anything, to be faintly *female*. If your reaction is like mine, it does. And as with the color illusion, you can also perform the opposite experiment. Fixate for at least ten seconds on the isolated hyperbolic female face, and then suddenly refixate on the neutral face to see if it subsequently looks, if only for a second or so, to be distinctly *male*.[3] Be your own judge. (You may also notice an intriguing *age* effect. The central face appears as two entirely *distinct individuals* in the two conditions. When seen as male, it looks quite young—no more than twenty. When seen as female, it looks

3. A more elaborate and carefully controlled experiment of this sort is reported by O'Toole and company (see Blanz et al. 2000). The distortions induced in normal face perception, by various hyperbole-induced profiles of neuronal fatigue, extend to selectively *facilitating* the recognition of specific test-faces. This happens when the test subject is 'primed' by face-images deliberately constructed to be face-space opposites, or '*anti*-faces,' of the face to be identified.

to be a much older person—no less than thirty-five. I'm still pondering that one.)[4]

There are many charming variations on these experimental themes, for both the color space and the face space, but we must here move on. That these spaces *exist*, that they display a determinate internal *structure*, and that they have an at least slightly *plastic* neuronal basis, are the suggestive lessons of this introductory discussion.

3   Individual Learning: Slow and Structural

Still, one might ask of these introductory spaces if they are perhaps *innate*— that is, somehow specified in the human genome. Whatever the correct answer here, it is surely empirical. As for the human color space, the answer may well be positive. As we shall see later in this book, the probable neuronal and synaptic basis for the three-dimensional color solid discussed above seems to be anatomically simple, repetitive, and highly uniform across normal human individuals. It is therefore a fair candidate for being somehow specified in the human genome.

On the innateness of the human representational-space for *faces*, however, our judgment must incline to the negative, for several reasons. First, and unlike the three-dimensionality of both your color space and the (schematic) face space of figure 1.2, the dimensionality of your brain's face space lies almost certainly in the thousands and perhaps in the tens of thousands. Second, the configuration of the synaptic connections that structure that face space must therefore include at least a million synaptic connections.[5] Unlike our color system, none of these connections are repetitive in strength or placement, so their genetic specification would be

---

4. These two examples portend a multitude. Readers may remember the three buckets of water—one tepid, flanked by one hot and one cold—touted by Locke and Berkeley. The tepid one feels warm, or cool, depending on which of the two flanking buckets initially fatigued the thermal receptors within the testing hand. The waterfall illusion provides a fourth example. Given three textured squares, a downflowing motion in the left square will eventually induce an illusory *upward* motion in the motionless central square, and an upward-flowing motion in the right square will induce the opposite illusion. What is intriguing about the color and face illusions, as distinct from the two illusions just cited, is that they occur in a representational space of more than one dimension—three in the case of colors and probably hundreds in the case of faces.

5. On average, each neuron in the brain enjoys perhaps 1,000 synaptic connections with other neurons.

costly. And third, those connections are not uniform across individuals in any case. We already know that the metrical structure of an individual person's face space *varies* substantially as a function of which culture she grew up in—specifically, as a function of which racial or ethnic group happened to exhaust or dominate her childhood experience with human faces. (It is sometimes called, inaccurately, the *other-race effect*.) In sum, the basis of our face space is complex, not recursively specifiable, and highly variable across individuals. It is therefore a most unlikely candidate for being coded in the genome.

Of course, the sheer *existence* of a neuronal population primed to take on the job of parsing faces is likely something that *is* genetically specified. After all, each of us has an inferotemporal cortex (the presumed region of the human brain that supports facial recognition), and all normal infants fixate on facelike stimuli from birth. But the adult structure of that space, its idiosyncratic dimensions, and its internal similarity-metric are all features that are *epigenetically* determined. By far the greater part, and perhaps even all, of what anyone knows about faces has its origins in one's *post*natal experience with faces.[6]

Such an antinativist conclusion, concerning the neuronal and synaptic basis of the brain's representational spaces, is appropriate for almost all of the many spaces that the brain comprehends. The only likely exceptions will be those occasional spaces—such as our color space, perhaps—that directly code the behavior of our various sensory neurons. There is a rule here that cannot be too far wrong. Specifically, the greater the *distance* between a given neuronal population and the body's sensory neurons—as measured by the number of distinct synaptic connections that have to be traversed for an axonal message to get from the one population to the other—the exponentially greater is the likelihood that the target population embodies a representational space that has been structured by *learning*.

The fact is, the modification, extinction, and growth of new *synaptic connections* is the single most dramatic dimension of structural change within the brain, from birth onward. Creating and adjusting the precious

6. One's prenatal *motor* and *proprioceptive* experience with one's own face, lips, and tongue may give one a leg up on the postnatal *visual* processing of postnatally encountered faces—for those 'visual' faces are homomorphic structures subject to homomorphic behaviors—but the 'knowledge' involved here is still epigenetic, even if acquired while still in the womb. The fact is, infants are highly active in the late stages of pregnancy: they move their tongues and mouths, and even suck their thumbs and fists.

configuration of one's 10^{14} synaptic connections is the very essence of learning in one's infant and childhood stages, for it is the collective configuration of the synaptic connections onto any neuronal population that *dictates* the family of categories embodied in that population's proprietary activation space. As our artificial neural-network models reveal, that collective configuration of synaptic connections is what *shapes* the web of similarity and difference relations that unite and divide the categories within the resulting activation space. And that same collective configuration of synaptic connections is what *transforms* any incoming activation pattern (from, for example, an earlier population of sensory neurons) into a new activation pattern located at a specific location within this carefully sculpted secondary space. Those assembled connections are thus central to *both* the relatively enduring conceptual framework that is acquired by the learning individual in the first place, *and* to its subsequent moment-to-moment activations by ephemeral sensory inputs. These synaptic connections are, simultaneously, the brain's elemental information *processors*, as well as its principal *repository of general information* about the world's abstract structure.

Accordingly, the establishment of one's myriad synaptic connections, and the fine-tuning of their individual strengths or 'weights,' constitutes the focal and primary process of learning that anyone's brain undergoes in its first twenty years of life, and especially in the first ten, and most especially in the first five. It is during these periods that one's background conceptual framework is slowly established, a framework that is likely, with only minor modifications, to remain with one for the rest of one's life.

Notably, and despite its primacy, that synapse-adjusting/space-shaping process is almost wholly ignored by the traditions of academic epistemology, even into these early years of our third millennium. This is perhaps not too surprising. The profound inaccessibility of the brain's microprocesses, the absence of compensatory computer models thereof, the focus of professional concern on normative matters, and the tyrannical primacy of folk-psychological conceptions of cognitive activity collectively contrive for *past* theorists an acceptable excuse for this monumental and crippling oversight. But the validation-dates on these several excuses (as in "use before 01/01/2001") have all run out.

Beyond conventional microscopy, an ever-growing armory of experimental techniques and instruments—such as the selective staining of neurons and their connecting pathways, electron microscopy, single-cell microelectrodes, multiple-array microelectrodes, genetically modified mice, CAT scans, PET scans, MRI scans, fMRI scans, activity-sensitive florescent dyes—now provides us with an overlapping set of windows onto the brain's physical structure and its neuronal activities, from the subcellular details of its molecular activities up to the molar-level behavior of its brainwide neuronal networks. The brain is no longer an inaccessible black box. On the contrary, the steady stream of experimental data just mentioned provides an ever-expanding set of empirical *constraints* on responsible cognitive theorizing. It also makes possible the invaluable sorts of back-and-forth interactions, between theoretical suggestions on the one hand and experimental probings on the other, that have proven so successful in other scientific endeavors. Theories can suggest and motivate experiments that might never have been tried, or even conceived, in their absence. Equally, experimental results can force needed modifications in the theories being tested. And the cycle begins anew.

Artificial neural networks, as realized directly in electronic hardwares, or as modeled at one remove in conventional computers, provide us with a second way of formulating and testing theories of how cognitive activities can arise from interconnected sets of neuron-like elements. Unlike biological networks, the artificial networks can be as simple and as well controlled as we wish. We can monitor their every connection, their every twitch, and their every change, without killing, damaging, or interfering with the network elements involved. Since, being electronic, they also run much faster than do their biological namesakes, we can also perform and complete experiments on their 'learning' activities within *hours*, experiments that would take months or years to perform on a biological brain. This means, of course, that we can often learn very swiftly that our initial attempt to model some brain function or other is just *wrong*. We can thus go back to the empirical brain for renewed theoretical inspiration, and then go back to the drawing board in hopes of constructing a more faithful artificial model.

This activity has already yielded the rough outlines of a competing conception of cognitive activity, an alternative to the "sentential" or "propositional-attitude" model that has dominated philosophy for the past 2,500 years. This newer conception is important for a number of reasons, not least because it finally (finally!) makes some explanatory contact with the physical and functional details of the biological organ—the brain—that actually sustains our cognitive activity. But there are many other reasons as well, perhaps the first of which is the novel account it provides of the *origins* of any individual's conceptual framework.

This is an issue that is mostly ignored, submerged, or finessed with a stick-figure theory by the philosophical tradition. Two styles of 'solution' have dominated. The writings of Plato, Descartes, and Fodor illustrate the first great option: since one has no idea how to explain the origin of our concepts, one simply pronounces them innate, and credits either a prior life, almighty God, or fifty million years of biological evolution for the actual lexicon of concepts we find in ourselves. The works of Aristotle, Locke, and Hume illustrate the second great option: point to a palette of what are taken to be sensory 'simples,' such as the various tastes, smells, colors, shapes, sounds, and so forth, and then explain our base population of simple concepts as being faint 'copies' of the simple sensory originals, copies acquired in a one-shot encounter with such originals. Nonsimple or 'complex' concepts are then explained as the recursively achieved concatenations and/or modulations of the simple 'copies' (and never mind the origins of *that* constructive machinery).

Both options are hopeless, and for interesting reasons. If we put Almighty God and Plato's Heaven aside as nonstarters, the preferred (i.e., the evolutionary) version of the first option confronts the difficulty of how to code for the individual connection-places and connection-strengths of fully 10^{14} synapses—so as to sculpt the target conceptual framework—using the resources of an evolved genome that contains only 20,000 genes, 99 percent of which (all but a paltry 300 of which) we share with *mice*, with whom we parted evolutionary company some fifty million years ago. The problem here is not so much the nine-orders-of-magnitude gap between the number of available genes and the number of synaptic connections (in principle, a recursive procedure can bridge a gap of any size), although that yawning gap does give pause for much thought. The real difficulty is the empirical fact that each person's matured synaptic configuration is radically different from anyone else's. It is utterly unique to that individual. That synaptic configuration is thus a hopeless candidate for being recursively specifiable as *the same* in all of us, as it must be if the numbers gap just noted is to be recursively bridged, and if the same conceptual framework is thereby to be genetically recreated in every normal human individual. (More on this later.)

The second option is little or no better than the first. Empirical research on the neuronal coding strategies deployed in our several sensory systems reveals that, even in response to the presumptively 'simplest' of sensory stimuli, the sensory messages sent to the brain are typically quite complex, and their synapse-transformed offspring—that is, the downstream conceptualized representations into which they get coded—are more complex still, typically *much* more complex. (More on this later, also.) The direct-inner-copy theory of what concepts are, and of how we acquire them, is a joke on its face, a fact reflected in the months and years it takes any human infant to acquire the full range of our discriminatory capacities for most of the so-called 'simple' sensory properties. (It takes *time* to configure the brain's 10^{14} synapses, even to comprehend the 'simple' properties at issue.) Additionally, and as anyone who ever pursued the matter was doomed to discover, the recursive-definitions story suggested for 'complex' concepts was a crashing failure in its own right. Try to construct an explicit definition of "electron," or "democracy"—or even "cat" or "pencil," for that matter—in terms of concepts that plausibly represent sensory simples.

Perhaps the strongest argument that either side—blanket nativism versus concatenative empiricism—could adduce in its own favor was the evident poverty of the opposing view. Given the depth of their poverties, each had a nontrivial point, so long as these alternatives were seen as exhausting the possibilities. But of course they do not. As we shall see in the next chapter, we already possess a workable story of how a neuronal activation space can be slowly sculpted, by experience, into a coherent and hierarchical family of prototype regions. This story also accounts for the subsequent context-appropriate *activation* of those concepts, activations made in response to sensory instances of the categories they represent. And, as we shall see in subsequent chapters, this same neurostructural and neurofunctional framework sustains penetrating explanations of a wide variety of perceptual and conceptual phenomena, including the profile of many of our cognitive *failures*. It forms the essential foundation of the larger epistemological theory to be outlined in the chapters to come.

4   Individual Learning: Fast and Dynamical

But *only* the foundation. Adjusting trillions of synaptic connections is not the only way to engage in a process worthy of the term "learning," and boasting a well-tuned configuration of such connections is not the only way to embody systematic 'knowledge' about the world. To this basic dimension of learning—the dimension of *structural* changes in the brain— we must add a second dimension of learning: the dimension of *dynamical* changes in the brain's typical or accustomed modes of operation. These dynamical changes can take place on a much shorter time scale (seconds and less) than the structural changes we have been discussing (weeks, months, and years), and they typically involve no structural changes whatever, at least in the short term. But this dimension of cognitive development is at least as important as its structural precursor, as we shall see.

One can conceive of the brain's dynamical activity as a single moving point in the brain's 'all-up' neuronal activation space, a point in ceaseless motion, a point that spends its time, marble-like, rolling merrily around the almost endless hills and valleys of the conceptual landscape that the basic or structural learning process has taken so long to sculpt. This landscape analogy is accurate enough, in that it rightly suggests that one's unfolding cognitive state tends to favor the 'valleys' (the acquired prototype regions or categories) and to slide off the 'hills' (the comparatively improbable ridges between the more probable valleys). But it fails adequately to suggest the spectacular *volume* of the brain's all-up activation space (the assembled sum of the thousands of vaguely Kantian *sub*spaces). Let us do a brief accounting.

If we assume, very conservatively, that each neuron in the brain admits of only ten different functionally significant levels of activity—ten steps between a minimum spiking frequency of 0 Hz and a maximum of 90 Hz, for example—then, since the brain has 10^{11} neurons, we are looking at a space of 10 to the 10^{11} power or $10^{100,000,000,000}$ functionally distinct, a priori possible global activation states. (For comparison, the accessible universe contains only about 10^{87} cubic meters.) It is within this almost incomprehensible volume of distinct activational possibilities that a given individual's moving activation-point must craft its one-shot, three-score-and-ten years, idiosyncratic cognitive excursion—that individual's conscious (and unconscious) life.

That space is far too large to explore any significant portion of it in a human lifetime ($\approx 2 \times 10^9$ seconds). If one's activation point were to streak through the possible points at a rate of 100 per second, a lifetime's exploration would touch only 2×10^{11} distinct points, leaving fully $5 \times 10^{99,999,999,988}$ points unvisited.

This accounting is roughly accurate, but it concerns only the a priori volume of the human neuronal activation space, that is, its potential volume if each neuron's activities were independent of every other's. By contrast, its *a posteriori* volume, if still mind-boggling, is substantially smaller, for a reason the reader will likely already appreciate. The whole point of the synapse-adjusting learning process discussed above was to make the behavior of neurons that are progressively higher in the information-processing hierarchy profoundly and systematically *dependent* on the activities of the neurons below them. Accordingly, that learning process shrinks the space of (empirically) possible global activation points dramatically.

More specifically, it shrinks the original space to a set of carefully contrived internal subspaces, each of which is an attempt to represent—via its acquired internal structure—some proprietary aspect or dimension of the external world's enduring structure. Being devotedly general in their representational significance, these subspaces therefore represent the brain's conception of the full range of *possible* ways in which the actual world may present itself to us in our singular, ongoing perceptual experience. For example, the color space of figure 1.3 attempts to represent the range of all possible colors. The face space of figure 1.4 attempts to represent the range of all possible human faces. A third space might represent the range of all possible projectile motions. A fourth, the range of all possible voices. And so on. Collectively, these many subspaces specify a set of 'nomically possible' worlds—worlds that instantiate the same categories and share the enduring causal structure of our own world, but differ in their initial conditions and ensuing singular details.

Accordingly, those spaces hold the key to a novel account of both the semantics *and* the epistemology of modal statements, and of counterfactual and subjunctive conditionals. The account envisioned here does not require us to be swallow-a-camel 'realists' about the objective existence of possible worlds, nor does it traffic in listlike state-descriptions. In fact, the relevant representations are entirely nondiscursive, and they occupy positions in a space that has a robust and built-in *probability metric* against which to measure the likelihood, or *un*likelihood, of the objective feature represented by that position's ever being instantiated.

Once again, the acquired microstructure of the brain holds out some much-needed explanatory resources, but the price of using them is a reconception of our modal, subjunctive, and counterfactual knowledge, a reconception that moves us away from the linguaformal framework of Folk Psychology and classical logical theory, and toward the activation-vector-space framework appropriate to the brain's basic modes of operation.

This shift in focus will be a gain in one further respect. Nonhuman animals demonstrate their possession of modal, subjunctive, and counterfactual knowledge no less than do humans. But nonhuman animals do not traffic in sentences or in propositional attitudes at all, let alone in their evaluation across an infinity of possible worlds. The classical accounts, therefore, cannot explain the great bulk of modal knowledge throughout the animal kingdom. They cannot get beyond the idiosyncratic case at which they were initially aimed, namely, language-using humans. The present approach, by contrast, holds out a uniform account, of broadly 'modal' knowledge, for all terrestrial creatures.

As stated earlier, one's inner cognitive narrative is a specific trajectory through such an antecedently sculpted space. But even given such a well-informed space, the path of that cognitive trajectory is not dictated by one's sensory experience alone. Far from it. At any point in time, your next activation point, within your global activation space, is always dictated (1) partly by your current sensory inputs, (2) partly by the already acquired profile of your background conceptual framework (that is, by the lasting configuration of your synaptic connections), but also, and most importantly in the present context, (3) *by the concurrent activation-state of your entire neuronal population*, a complex factor that reflects your cognitive activity immediately preceding the present computational interaction. This arrangement makes the brain a genuine *dynamical system*, a system capable of a great range of possible behaviors, much of which is unpredictable even in principle.

Once again, a Kantian parallel may help to illuminate the claims here being made, by the contrasts required as well as by the similarities in place. It is arguable that Kant embraced a third great canvas of cognitive activity, distinct from Intuition and Judgment. This was the faculty of Imagination, whose hallmark feature was the *spontaneity* of the cognitive activities there displayed. Unlike Kant, I shall not postulate a distinct space or canvas to sustain the activities of the imagination. On my view, those activities take place in the very same family of neuronal activation spaces discussed earlier. What distinguishes imaginative activity from other forms of cognition is not its location, but its cause. Imaginative cognitive activity arises not from ascending inputs from the several sensory modalities, but from descending or recurrent inputs from neuronal populations higher up in the brain's information-processing hierarchy. It is initiated and steered by brain activity from above rather than by sensory activity from below.

On the matter of spontaneity, however, I line up with Kant. And with good reason. The brain is a dynamical system of unparalleled complexity. It is a continuously varying physical system with many billions of degrees of freedom—the current activation levels of its billions of neurons, for starters. It is a system whose dynamics are decidedly nonlinear, which means that, for many regimes of activity, infinitesimally small differences in one's current cognitive state can snowball exponentially into very large differences in the brain's subsequent cognitive state. This puts inescapable limitations on the degree to which we, or any other conceivable physical device for that matter, can *predict* any brain's unfolding cognitive activities, even on the assumption that the brain's behavior is rigorously deterministic. The problem is that, for such a system, effective prediction requires, first, infinitely perfect information about the brain's current structural and dynamical state, and second, infinitely accurate computations concerning its law-governed development into subsequent states. Neither requirement can be met in this world, nor even relevantly approximated.

The result is a system whose cognitive behavior, in general, simply cannot be predicted—not by itself, and not by anything else either. This need not mean that *no* regularities will display themselves to the casual eye. To the contrary, when the brain is in the midst of some prototypical activity, such as brushing one's teeth, dealing a deck of cards, or sipping a cup of coffee, its specific motor behavior can be reliably predicted for several seconds or so into the future. And if we look at behavior as reckoned over days and weeks, we can reliably predict that, if the environment remains normal, people will take dinner about six o'clock, go to bed around ten, and get up around six or seven. The details of such periodic behaviors may be beyond us (Will he have sausages, or fish? Will he wear the green pajamas, or the blue? Will he put on the right shoe first, or the left?) But even a nonlinear system can display roughly stable, if endlessly variant, orbits or cycles. Beyond these two exceptions, however (very short-term behaviors, and long-term patterns), a person's cognitive and motor behavior is deeply unpredictable. It displays a spontaneity that reflects its origins in an unmonitorable and ever-shifting mix of mercurial microprocesses.

But let us return to the topic of learning. Beyond a welcome measure of spontaneity, what the recurrent or descending neuronal pathways also make possible is the ongoing *modulation* of the brain's cognitive response to its unfolding sensory inputs. The details of that modulation reflect the ever-changing dynamical state of the brain as a whole; it reflects all of the acquired here-and-now contextual information embodied in the brain at the time of the sensory inputs in question. Most importantly, the current context into which each sensory input arrives is never exactly the same twice, not twice in a lifetime. For even with the brain's *synaptic* connection-weights fixed in their mature configuration, the brain's *dynamical* state—its all-up pattern of *current neuronal activation-levels*—provides an ever-moving, never-repeating cognitive context into which its every sensory input is interpretively received. Accordingly, one never has a truly identical cognitive response on two different occasions, even if the total sensory inputs should happen to be identical on those two occasions. Identical rocks, thrown sequentially into this internal Heraclitean river, will never make exactly the same splash.

To be sure, the resulting differences are usually small, and their downstream cognitive consequences are typically small as well. The brain, like the solar system, is at least a quasi-stable dynamical system. But sometimes the downstream differences are substantial, and reflect a changed outlook on the world, as when a trusted friend suddenly mistreats an innocent colleague horribly. Though the friend may return to normal in the days and weeks that follow, his smiles and greetings and other social exchanges never look quite the same to you again. Your perceptions, anticipations, and interactions, at least where he is concerned, are permanently altered: you have learned something about his character.

More specifically, that experience has kicked your cognitive trajectory into an importantly different and hitherto unvisited region of your antecedently sculpted neuronal activation space. In the vast majority of that space's many dimensions, your trajectory remains in familiar territory. But in at least a handful of the available dimensions, that trajectory is now exploring new ground.

Strictly, it should be said, that trajectory is *always* exploring novel territory, since it never intersects itself *exactly*. (For an isolated dynamical system, such a perfect return would doom the system to an endless and unchanging periodicity.) But sometimes the novelty of one's activation-space position is not minor: it is substantial. Occasionally, one's redirected trajectory takes one out of a familiar and much-toured basin of dynamical attraction, over a local ridge of relative improbability, and into a new and quite different basin of attraction, a basin in which *all* sensory inputs of a certain kind now receive an importantly different regime of conceptual interpretation. If that new regime happens to yield an increased capacity for anticipating and manipulating one's environment, or some specific aspect of it, then it is appropriate to credit the creature with a new insight into the world. Though no *structural* changes have taken place in one's nervous system, such a case is still a clear instance of learning—we may call it *dynamical* learning.

The example of the misapprehended and then reapprehended friend is a deliberately mundane example of the process. That process will loom larger in importance when one appreciates that most cases of major scientific insight or so-called 'conceptual revolution' are also instances of dynamical learning. Consider, for example, Newton's famously sudden realization (the falling apple incident at Woolsthorp) that the Moon's orbit is just another case of *projectile motion*, governed by the same laws as a stone thrown here on Earth. Consider Darwin's realization that the origin of distinct species might be owed to an entirely natural analogue of the *artificial selection* long practiced in animal husbandry. Consider Torricelli's insight that we are all living at the bottom of an *ocean of air*, and the test of that hypothesis with a (steadily falling) barometer lugged manfully up a mountainside. Or consider Bernoulli's, Maxwell's, and Boltzmann's conjecture that a gas is just a cloud of *tiny ballistic particles* in rebounding collision with each other, and with the walls of whatever container might confine them.

In all of these cases, and in many others, the initial cognitive change effected lay not in the reconfiguration of anyone's synaptic weights—the changes at issue happened much too swiftly for that molasses-like process to provide the explanation. Rather, the change consisted in the *dynamical redeployment* of conceptual resources already in place, resources learned years ago and in other contexts entirely, resources *originally* learned by way of that slower synaptic process here found wanting. What was novel in the historical examples cited above was not the concepts deployed (*inertial projectile, selective reproduction, deep ocean*, and *swarm of ballistic particles*). These concepts were already in place. The novelty lay in the unusual target or circumstance of their deployment—namely, the *Moon, wild animals*, the *atmosphere*, and *confined gases*, respectively. In each case, an old and familiar thing came to be understood as an unexpected instance of a quite different category, a category hitherto employed in quite different circumstances, a category that makes new and systematic sense of the old phenomenon. To borrow a notion from biology, we are here looking at a variety of *cognitive exaptations*—that is, at cognitive devices initially developed in one environment that turn out to serve unusually well for a different purpose in a different environment.

As the reader may begin to surmise, the account of dynamical learning to be outlined in chapter 4 offers a new account of what theoretical hypotheses are, of what explanatory understanding consists in, and of what explanatory unification or "intertheoretic reduction" consists in. Its classical adversaries are the *syntactic* account of theories ("a theory is a set of sentences"), with its appropriately *deductive* accounts of both explanation and intertheoretic reduction, and the *semantic* view of theories ("a theory is a family of homomorphic models"), with its appropriately *model-theoretic* accounts of explanation and reduction. Both of these classical accounts are inadequate, I shall argue, especially the older syntactic/sentential/propositional account. Among many other defects, it denies any theoretical understanding whatever to nonhuman animals, since they do not traffic in sentential or propositional attitudes. That classical account is deeply inadequate for humans also, since it wrongly attempts to apply *linguistic* categories, appropriate primarily at the social level, to the predominantly *non*linguistic activities of individual brains. These primordial activities are better described in terms of an *unfolding sequence*

of activation-vectors than in the procrustean terms of sentential or propositional attitudes. There is ample room for sentential representations within
the epistemological story to be presented below, and important work for
them to do. But their proper home lies in the social world—in the shared
space outside of the human brain, in the space of public utterances and
the printed page—not inside the individual human head.

The semantic view of theories is mistaken also, but of these two classical
adversaries, it comes much closer to the truth of the matter, as we shall see
in due course. Despite its esoteric and, to many, rather daunting standard
formulation in set-theoretic terms, a specific theory can also be relevantly
characterized, on the semantic view, as a *single complex predicate*, a predicate
that may be true of a specified domain-to-be-explained.[7] Thus, the Moon
is a *Newtonian projectile*; the evolution of animal species is a *process shaped
by random variation and selective reproduction*; the particles of a gas constitute
a *classical mechanical system with perfectly elastic collisions*; and so on. (This
predicative characterization misleads slightly, since it returns us to linguistic items such as predicates. By contrast, the canonical formulation regards
the theory proper as being an *extensional* entity—the set-of-all-models of
which that predicate is true. But never mind.) What the *neurosemantic*
account of theories does is rather similar to the semantic account, although
it locates theories firmly inside the head. Simply replace "single complex
predicate" with "single prototype point in high-dimensional activation
space," and you have the outlines of the view to be defended below.

The semantic view of theories is still a minority view, however, even
among epistemologists, and so the reader may find the parallel just drawn
to be neither illuminating nor very compelling. Let me therefore draw
another parallel, which comes closer still to the truth of the matter. The
view of theories to be defended here is a neurally grounded instance of the
tradition represented by Mary Hesse, Thomas Kuhn, Ronald Giere, William
Bechtel, Nancy Cartwright, and Nancy Nersessian. That tradition focuses
on the role of models, metaphors, paradigms, mechanisms, and idealized
'nomological machines' in the business of scientific theorizing.[8]

7. Cf. van Fraassen 1980.
8. This list is confined specifically to philosophers of science, but the theoretical
tradition here drawn upon reaches beyond the confines of that particular subdiscipline. 'Cognitive Linguistics,' as represented by the work of linguists such as J. Elman,
E. Bates, R. Langacker, G. Lakoff, and G. Fauconnier, is a different but salient branch
of the same tree. 'Semantic-Field Theory,' as represented by the work of philosophers
such as E. Kittay and M. Johnson, is another. And so is the 'Prototype Theory' of
conceptual organization explored by an entire generation of psychologists.

In some respects, this is a highly diverse group. Famously, Kuhn's focus lay at the social level; Giere's and Nersessian's focus lies firmly at the psychological level; Bechtel is himself an occasional neuromodeler; and Cartwright tends to focus on the metaphysical nature of objective reality. But they are all united in seeing our scientific endeavors as dependent on the artful *assimilation* of complex and problematic phenomena to some special phenomena that are familiar, tractable, and already well understood. It is the neurocomputational *basis* of precisely such assimilative processes that I aim to capture as instances of dynamical learning. In the chapters that follow, we shall explore the *consequences* of that capture for a range of standard issues in epistemology and the philosophy of science.

One may still hear echoes of the semantic account of theories in the above list. After all, Giere counts his own view as one (heterodox) instance of that account, and Sneed (1971) and Stegmuller (1976) have tried to put Kuhn's story into model-theoretic dress, with perhaps some success. Well and good. I own a residual sympathy for the semantic account, alien though it may seem. And I agree with the first part of van Fraassen's (1980) important claim that the proper medium for representing the essence of scientific theories is not the formalism of classical *first-order logic*, but rather the formalism of *set theory*. My disagreement concerns the second half of his claim. The proper formalism, I suggest, is not set theory either (or not *only* set theory). The proper formalism is *vector algebra* and *high-dimensional geometry*. And the proper deployment of this formalism is the story of how vector coding and vector processing are realized in the vast network of the biological brain. This allows us, among other things, to bring a *dynamical* dimension into our account of human scientific theorizing—it is, after all, a causal process unfolding in time—a dimension largely or wholly absent from the original semantic accounts.

Before leaving this introductory discussion of dynamical or second-level learning, it is worth mentioning that one further major problem in the epistemological tradition is going to show up in a new and potentially more tractable guise, namely, the problem of the underdetermination of theory by evidence, and the status of broadly Realist versus broadly Instrumentalist interpretations of the enterprise of science. The underdetermination problem does not disappear—far from it—but it does assume a different form, and it heralds, I shall argue, a philosophical lesson somewhat different from that urged by either of these traditional adversaries. For one thing, as the reader may appreciate, the 'evidence' relation needs to be reconceived entirely, since the parties to it are no longer the 'theoretical' and the 'observational' sentences of the syntactic view, nor the set-theoretic structures and their 'observational substructures' embraced by the semantic view. For another, we are going to find that underdetermination infects the domain of all possible *evidence* no less than the domain of all possible theories, with consequences we shall have to evaluate. And for a third, we shall rediscover another old friend— incommensurability (also reconfigured)—as we confront the practical infinity of neurocognitive alternatives potentially vying for human accep- tance as the preferred vehicle of our global understanding. Despite these familiar bogeymen, it will be a recognizable version of Scientific Realism, as I see it, that best makes sense of the overall situation.

5 Collective Learning and Cultural Transmission

If it is important to distinguish a brain's *dynamical* adventures (in trying to apply its existing concepts to an ever-expanding experience of the world) from its more basic *structural* adventures (in slowly shaping a useful framework of concepts in the first place), then it is equally important to distinguish both of these originally individual activities from a third major level of learning—the level of *cultural* change and *collective* cognitive activ- ity. For it is the institution of this third level of learning that most surely distinguishes the cognitive adventure of humans from that of any other species. This third-level activity consists in the cultural assimilation of individual cognitive successes, the technological exploitation of those suc- cesses, the transmission of those acquired successes to subsequent genera- tions, and the ever-more-sophisticated *regulation* of individual cognitive activities at the *first two* levels of learning.

The existence and overwhelming importance of this third level of cogni- tive activity will be news to no one. But the proper characterization of that collectivized process is still a matter of substantial dispute. Is it the journey of geist toward complete self-consciousness, as Georg Hegel surmised? Is it the reprise of selective evolution at the level of linguistic items—a ruthless contest of selfish 'memes'—as Richard Dawkins has suggested? Is it the convergent march of science toward the Final True Theory, as some Prag- matists and Logical Empiricists dared to hope? Is it just the meandering and ultimately meaningless conflict between fluid academic fiefdoms com- peting for journal space and grant money, as some skeptical sociologists have proposed?

A wry answer might be that it is all of these. But a more considered and more accurate answer would be that it is none of the above. Nor will the true nature of this third-level process ever become clear until we appreciate the manifold ways in which the various mechanisms of human culture serve to nurture, to regulate, and to *amplify* the cognitive activities of individual humans at the first two levels of learning, the levels we share with nonhuman animals.

As the offhand list of the preceding paragraph will attest, there is no shortage of philosophical theories about the structure, dynamics, and long-term future of cultural or third-level learning. They are many and various. But if the proposal here on the table—that the central *function* of these cultural mechanisms is the detailed exploitation and regulation of learning at the first two levels—then none of the familiar theories can hope to be anything more than incidentally or accidentally correct in their portrayals of the human epistemic adventure. For no epistemology or philosophy of science prior to the present period has had any interest in, or any clear conception of, these first two kinds of learning—namely, the generation of a hierarchy of prototype-representations via gradual change in the configuration of one's synaptic weights (first-level or 'structural' learning), and the subsequent discovery of successful redeployments of that hard-earned framework of activation-space representations, within novel domains of experience (second-level or 'dynamical' learning).

Indeed, those original and more basic levels of representation and learning have been positively *mis*characterized by their chronic portrayal as just hidden, inward versions of the *linguistic* representations and activities so characteristic of cognitive activity at the third level. As noted earlier, Jerry Fodor is the lucid, forthright, and prototype perpetrator on this particular score, for his theory of cognitive activity is that it is explicitly language-like from its inception (see, e.g., Fodor 1975)—a view that fails to capture anything of the very different, *sub*linguistic styles of representation and computation revealed to us by the empirical neurosciences and by artificial neuromodeling. Those styles go wholly unacknowledged. This would be failure enough. However, the 'Language of Thought' hypothesis fails in a second monumental respect, this time, ironically enough, by *under*valuing the importance of language. Specifically, it fails to acknowledge the extraordinary cognitive *novelty* that the invention of language represents, and the degree to which it has launched humankind on an intellectual trajectory that is impossible for creatures denied the benefits of that innovation, that is, for creatures confined to only the first and second levels of learning.

What I have in mind here is the following. With the emergence of language, the human race acquired a public medium that embodied—in its peculiar lexicon and in its accepted sentences—at least some of the acquired wisdom and conceptual understanding of the adults who share the use of that medium. Not *all* of that acquired wisdom. Not by a long shot. But enough of it to provide an informed template to which the conceptual development and dynamical cognition of subsequent generations could be made to conform. These subsequent generations of language-learners and language-users are thus the heirs and beneficiaries of at least some of the cognitive achievements of their forebears. In particular, they do not have to sculpt a conceptual space entirely from scratch, as nonlinguistic animals do, and as prelinguistic humans must have. To the contrary, as human children learn their language, from their parents and from the surrounding community of conceptually competent adults, they can shape their individual conceptual developments to conform, at least roughly, to a hierarchy of categories that has already been proven pragmatically successful by a prior generation of cognitive agents.

At that point, the learning process is no longer limited to what a single individual can learn in a single lifetime. That collective medium of representation—language—can come to embody the occasional cognitive innovations of many different human individuals, and it can accumulate those innovations over hundreds and thousands of lifetimes. Most importantly, the conceptual template that the language embodies can slowly *evolve*, over historical periods, to express a different and more powerful view of the world than was expressed by its more primitive precursors.

It is important not to overstate this point. Almost *all* of anyone's acquired wisdom goes with him to the grave, including his inevitably idiosyncratic command of the resources of human language. There is no realistic hope of recording the specific configuration of anyone's 10^{14} synaptic weights, and no realistic hope of tracing the dynamical history of anyone's brainwide neuronal activations, and thus no realistic hope of recreating, exactly, one's current brain state within the skull of another human. But one can hope to leave behind at least something of one's acquired understanding, if only a faint and partial digest thereof, through the communities of conversation and shared conceptual practice that one's speech behavior—whether live or printed—has helped to shape.

This said, it is equally important not to *under*state the importance of language. As a public institution whose current lexicon, grammar, and network of broadly accepted sentences are under no individual's exclusive personal control, a living language thereby constitutes a sort of 'center of cognitive gravity' around which individual cognitive activities may carve out their idiosyncratic but safely stable orbits. Moreover, as a cultural institution that long outlives the ephemeral individual cognizers that sequentially pass through it, a language embodies the incrementally added wisdom of the many generations who have inevitably reshaped it, if only in small ways, during the brief period in which it was theirs. In the long run, accordingly, that institution can aspire to an informed structure of categories and conventional wisdom that dwarfs the level of cognitive achievement possible for any creature living outside of that transgenerational framework. Large-scale *conceptual* evolution is now both possible and probable.

Everyone will agree, of course, that a species with some mechanisms for historical recording can achieve more than a species with no such mechanisms. But I am here making a rather more contentious claim, as will be seen by drawing a further contrast with Fodor's picture of human cognition. On the Language of Thought (LoT) hypothesis, the lexicon of any *public* language inherits its meanings directly from the meanings of the innate concepts of each individual's innate LoT. Those concepts derive *their* meanings, in turn, from the innate set of causal sensitivities they bear to various 'detectable' features of the environment. And finally, those causal sensitivities are fixed in the human genome, according to this view, having been shaped by many millions of years of biological evolution. Accordingly, every normal human, at whatever stage of cultural evolution, is doomed to share the *same* conceptual framework as any other human, a framework that the current public language is therefore secondarily doomed to reflect. Cultural evolution may therefore *add* to that genetic heritage, perhaps considerably, but it cannot undermine it or supersede it. The primary core of our comprehensive conception of the world is firmly nailed to the human genome, and it will not change until that genome is changed.

I disagree. The lexicon of a public language gets its meanings not from its reflection of an innate LoT, but from the framework of broadly accepted or culturally entrenched sentences in which they figure, and by the patterns of inferential behavior made normative thereby. Indeed, the sublinguistic categories that structure any individual's thought processes are shaped, to a significant degree, by the official structure of the ambient language in which she was raised, not the other way around.

To raise an even deeper complaint, the meaning or semantic content of one's personal cognitive categories, whether innate or otherwise, derives not from any feature-indicating nomic relations that they may bear to the external world. Rather, it derives from their determinate place in a high-dimensional neuronal activation-space, a space of intricate and idiosyncratic similarity relations, a space that embodies a highly informed 'map' of some external domain of lasting properties. As we shall see in the next chapter, the correct account of first-level learning requires us to put aside any form of atomistic, externalist, *indicator*-semantics in favor of a decidedly holistic, internalist, *domain-portrayal* semantics.

This means that both the semantic content of public languages and the semantic content of individual conceptual frameworks are not in the least bit 'nailed' to a fixed human genome. Both are free to vary widely, as a function of local epistemic circumstance and our individual and collective cognitive histories. But whereas each person's acquired conceptual framework is doomed to wink out after roughly three-score-and-ten years, the off-loaded, public structures of one's then-current language are fated to live on, in pursuit of an epistemic adventure that has no visible limits. Certainly this *third*-level world-representing process is not required to cleave to some Paleolithic or pre-Paleolithic conceptual framework somehow dictated by the human genome.

On the contrary, and given time, this third-level process opens the door to systematic reconstructions of our practical undertakings, and to systematic reconceptions of even the most mundane aspects of our practical and perceptual worlds. We can put out an initial public offering on the NYSE to support a company's plan to build a nuclear power plant to smelt aluminum from bauxite with megawatt applications of electric power. We can start a phone campaign to marshal registered Democrats to vote against the anti-choice initiative, Proposition 14, on the ballot of this year's state elections. We can aspire to write a thirty-two-bar tune in the key of G, based on the same chord sequence, appropriately transposed, as Harold Arlen's popular hit, "Stormy Weather." We can destroy the bacteria in a foul dishwashing sponge by boiling its residual water with klystron-generated microwaves in the countertop oven. We can marvel as we watch the spherical Earth rotate at fully fifteen degrees per hour, on its north–south axis, as the Sun 'sets' at the western horizon and the full Moon simultaneously 'rises' at the eastern horizon. Thoughts and undertakings such as these are simply beyond the conceptual resources of a Stone Age community of human hunter-gatherers. On the other hand, being the beneficiaries of their own third-level history, that Stone Age group has thoughts and undertakings—concerning such things as fire manipulation, food preparation, weapons technology, and clothing manufacture—that are equally inconceivable for the members of a baboon troop. We differ from the Stone Age humans by being at a very different rung on the long-term conceptual ladder that the institution of language provides. The baboons differ from us both, in having no such ladder to climb.

The institution of language is, however, only the first of many powerful mechanisms at this third and supraindividual level of learning. My hope, in the last chapter of this book, is to provide a novel perspective on all of them by exploring their roles in the regulation, amplification, and transmission of human cognitive activities at the first two levels of learning, as conceived within the neurostructural and neurodynamical frameworks outlined above. The payoff, if there is one, lies in the multilevel coherence of the portrait of human cognition that slowly emerges, and in the fertility of that portrait in grounding novel explanations of various aspects of human cognitive activity found problematic by our existing epistemological tradition. In short, I shall try to tell a new story about some old problems.

6 Knowledge: Is It True, Justified Belief?

The general motivation behind my rejection of the "true, justified belief" conception of knowledge should be roughly apparent already, even from the discussion of this introductory chapter. Let us examine each element of that wrongly probative analysis.

First, reflect on the required vehicle itself: the belief. This is an overtly *propositional* attitude, which requires for its specification a *declarative sentence*, a unit of representation not encountered in human cognition at least until the innovation of spoken language. It is not an element of either the first-level or the second-level processes of learning, as characterized above. To restrict knowledge to the domain of literal belief is therefore to deny any factual knowledge at all to animals and to prelinguistic human children, whose cognitive activities do not, or do not yet, involve the representational structures peculiar to language—not externally, and not internally, either.

Such a narrow conception will also ignore or misrepresent the great bulk of any *adult* human's factual knowledge, which reaches substantially beyond what little she can articulate in the comparatively clumsy vehicle of a declarative sentence. The relative poverty of such vehicles shows itself rather dramatically in the unexpected failures of the classical or program-writing research agenda within the field of artificial intelligence. A genuinely successful artificial intelligence, within any domain, requires a substantial knowledge base on which to ground its cognitive activities. But a long list of 'accepted sentences' proved to be a hopelessly inefficient means of storing information, and in addition, an exponentially *expanding* problem when we confront the task of retrieving the relevant bits of information as they are (unpredictably) needed. If you thought that the belief is, or deserves to be, the basic unit of factual knowledge, you had best think again.

Consider now the requirement of justification. That there is something awkwardly procrustean about this venerable condition becomes evident as soon as one considers again the case of factual knowledge in animals and prelinguistic human children. The fact is, the practice of justifying one's cognitive commitments is a practice that arises only among adult humans, only at the social level, and it typically involves a multipersonal negotiation concerning the status of some declarative sentence or other. One can and does rehearse such negotiations privately (and a good thing, too). But such rehearsals are confined to adult humans, and they reflect a practice whose origins and guiding prototypes lie in the public domain even there. In all, justification is a business that has its natural home at the social or third level of human learning, and its focus, naturally enough, is on representational vehicles—declarative sentences—that we have already found unequal to the task of being the basic units of human and animal cognition. I hasten to add that I have no desire in this book to denigrate the importance of our public institutions for the evaluation and justification of factual claims. Quite the contrary. But the "justified-true-belief" formula at issue wrongly imposes a condition that is drawn from and appropriate to a third-level learning process as a condition on knowledge *in general*.[9] It is wrong because the justification requirement has no application whatever to first-level knowledge, and only problematic or metaphorical application to the results of second-level learning. How do I go about 'justifying' to you, or even to myself, a specific trajectory just carved out in my neuronal activation space? Not in any usual way. Especially when those spontaneous trajectories are only feebly under my control in the first place.

9. This second condition has proven awkward even in advance of the more fundamental complaints voiced above. For example, one's unfolding perceptual beliefs typically count as knowledge, but the requirement that each one of them be *justified*, second by second, is difficult even to make sense of. Alvin Goldman (1986) has proposed, plausibly, that this traditional condition should therefore be replaced by the weaker but more apt requirement that the belief at issue have been produced by a "generally reliable mechanism." This is a step forward, but only a step. My complaint about the vehicles themselves remains unaddressed. And one wonders about those occasional cases where the generally reliable mechanism positively *malfunctions*, but, by the sheerest accident, happens to produce a true belief anyway. This meets the new condition, but it doesn't look like knowledge.

Finally, what about the requirement of truth? Surely, we are on firm ground in demanding truth as a condition on knowledge, at least if we restrict our attention to factual knowledge, as opposed to procedural knowledge or motor skills. But perhaps that ground is not so firm. An opening response to this plea for truth is that the neurocomputational perspective here to be explored makes it unexpectedly difficult to draw any clear or principled distinction between these two superficially different kinds of knowledge—factual knowledge versus practical skills—the one kind, truth-bearing; the other kind, not. A distinction that initially seems plain to uninformed common sense tends to disappear entirely when one confronts the realities of neuronal coding and world navigation, as we shall see.

A second and equally pregnant response concerns the sense of "truth" we are here supposed to impose. In its primary and literal sense it is a feature, once more, of declarative sentences. And its canonical explication in logical theory, the model-theoretic account of Alfred Tarski, leans entirely on the structural elements and grammatical features of human language.[10] That is an explication we have all learned to admire. But what are we supposed to say of unfamiliar representational vehicles such as a hierarchically structured, high-dimensional activation space—as a *conceptual framework* turns out to *be*, at bottom? And what are we supposed to say of representational vehicles such as a 10^6-element neuronal activation vector, which is what a *perceptual representation* turns out to be? These vehicles are not the sorts of things to which the notion of Tarskian truth even applies. And yet, as we have seen, they constitute the bulk of any creature's acquired factual knowledge. No doubt they can have, or lack, sundry representational virtues. But Tarski-style truth would seem not to be among them.

The "justified-true-belief" formula has, of course, received extensive critical attention in the literature. But the worries have focused almost exclusively on Gettier's (1963) illustration that these conditions are not yet collectively *sufficient* for knowledge. Most of the resulting research has bent its efforts in search of a *fourth* condition to patch up the perceived hole.

The complaint leveled here is different and deeper. It is that not one of these three familiar conditions is even individually *necessary* for knowledge. Indeed, the "justified-true-belief" approach is misconceived from the

10. Strictly, it leans on the structural features of a very tiny and highly regimented *fragment* of human language, namely, the narrow formalism of the first-order predicate calculus. But the account is still salutary. Let us give credit where credit is due.

outset, since it attempts to make concepts that are appropriate only at the level of cultural or language-based learning do the job of characterizing cognitive achievements that lie predominantly at the *sub*linguistic level. There are indeed dimensions of success and failure that operate at these more basic levels. But their articulation will require that we temporarily put aside the peculiar structures of language and address those deeper forms of representation on their own terms. This need not require that we simply toss away the notion of truth, as some may fear. But it will require that we reconceive it, and generalize it, in some highly specific ways. Knowledge, then, is *not* justified true belief, at least, not in general. As we shall see in the chapters to follow, it is something rather more interesting than that, and a good deal more precious.

Before closing this introduction, let me try to summarize. We are looking at three distinct learning processes at three quite different levels. At the first level, we find a process of structural change, primarily in the micro-configuration of the brain's 10^{14} synaptic connections. The product of this process is the metrical deformation and reformation of the space of possible activation patterns across each receiving neuronal population. This product is a configuration of *attractor* regions, a family of *prototype* representations, a hierarchy of *categories*—in short, a *conceptual framework*. In biological creatures, the time-scale of this unfolding process is no less than weeks, and more likely months, years, and even decades. It is slow, even molasses-like. It never ceases entirely, but by adulthood the process is largely over and one is more or less stuck with the framework it has created.

At the second level, we find a process of dynamical change in one's typical or habitual modes of neuronal activation, change that is driven not by any underlying synaptic changes, but by the recent activational history and current activational state of the brain's all-up neuronal population. Bluntly, the brain's neural activities are *self-modulating* in real time, thanks to the recurrent or feed-backward architecture of so many of its axonal projections. The product of this process is the systematic redeployment, into novel domains of experience, of concepts originally learned (by the first or basic-level process discussed above) in a quite different domain of experience. It involves the new use of old resources. In each case, this product amounts to the explanatory reinterpretation of a problematic domain of phenomena, a reinterpretation that is subject to subsequent evaluation, articulation, and possible rejection in favor of competing reinterpretations. The time-scale of such redeployments (in contrast to their subsequent evaluation and development) is much shorter than that of structural or basic-level learning—typically in the region of milliseconds to hours. This is the process that comprehends sudden gestalt shifts, 'eureka' effects, and presumptive conceptual revolutions in an individual's cognitive take on some domain of phenomena.

At the third level, we find a process of cultural change, change in such things as the language and vocabulary of the community involved, its modes of education and cultural transmission generally, its institutions of research and technology, and its techniques of individual and collective *evaluation* of the conceptual novelties produced at the first two levels of learning. This is the process that most decisively distinguishes human cognition from that of all other species, for the accumulation of knowledge at this level has a time-scale of decades to many thousands, even hundreds of thousands, of years. Its principal function is the ongoing regulation of the individual and collective cognitive activity of human brains at the first two levels of learning.

A cautionary note should be entered right here in the opening chapter. A close examination of the processes at all three levels will reveal a fractionation into sundry subprocesses and substructures. Most obviously, the third-level process is going to display a hierarchy of interlocking regulatory mechanisms, depending on the particular human culture, and on the particular stage of its history, that we choose to examine. Less obviously, but no less surely, the first-level process is a knit of architectural, developmental, neurochemical, and electrophysiological activities, which neuroscientists and neuromodelers have only begun to unravel. Similarly, the dynamical or second-level process will reveal its own variations and intricacies. The tripartite division sketched above is a deliberate idealization whose function is to provide a stable framework for more detailed exposition at each level, and for an exploration of the major interactions between the three levels.

If plainly acknowledged, there need be no deception in this. Useful idealizations are the life-blood of scientific understanding. What remains to be determined is: just *how* useful is the idealization at issue? Without more ado, let us now try to find out.

2 First-Level Learning, Part 1: Structural Changes in the Brain and the Development of Lasting Conceptual Frameworks

1 The Basic Organization of the Information-Processing Brain

Where to begin? At the bottom of the ladder. Or rather, at the bottom of several ladders, each of whose bottom rung is a population of *sensory* neurons—such as the rods and cones in the retina of the eye, the 'hair cells' in the cochlea of the ear, or the mechano-receptors within the skin (see fig. 2.1). Each of these initial neuronal rungs projects a wealth of filamentary axonal fibers 'upward' to a proprietary second rung—to a receiving population of postsensory neurons such as those in the lateral geniculate nucleus (LGN) for the *visual* system, to those in the medial geniculate nucleus (MGN) for the *auditory* system, to those in the central spinal cord for the widely distributed touch receptors of the *somatosensory* system, and so on.

The metaphor of a ladder, or a set of ladders, is entirely appropriate here. The neurons in each second rung project their axonal fibers in turn to a third rung, and that to a fourth and fifth, and so on up the processing hierarchy until we are ten, twenty, or even more projective steps above the initial sensory populations. Such ladders, at least in their lower stages, constitute the *primary pathways* for the visual system, the auditory system, the somatosensory system, and so forth. And each *rung* of each of these ladders constitutes a unique cognitive canvas or representational space, a canvas or space with its own structured family of categories, its own set of similarity and difference relations, and its own peculiar take on some enduring aspect of the external world. What happens, as sensory information ascends such a ladder, is its progressive transformation into a succession of distinct representational formats, formats that embody the brain's background 'expectations' concerning the possible ways-in-which the world can be.

Just how that multitracked hierarchy of 'expectations' is *constituted*, and how the brain *acquires* it, is the principal subject of this chapter. Here we

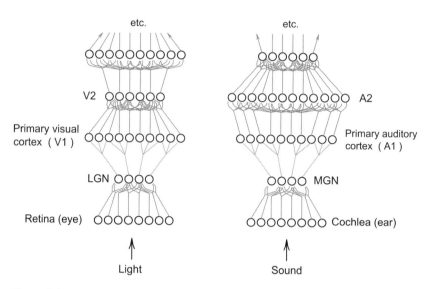

Figure 2.1
The ladder-like structure of the primary visual and auditory pathways.

need to focus on the synaptic interface that both separates and joins the neurons in any given rung from the neurons in the next rung up the processing ladder. The filamentary axons that reach up to that next rung typically divide, at their ends, into a large number of tiny axonal end-branches. (Think of an upraised arm branching at the hand into a number of upward-reaching fingers.) Each of those terminal fingertips makes a quasi-permanent *synaptic connection* with some neuron at the target population, a connection that permits the occasional flow of *neurotransmitter* molecules *from* that synaptic fingertip *to* the receiving cell. Such synaptic transmission, when it occurs, has the effect of either exciting the receiving neuron, if the connection at issue is of the positive or *excitatory* variety, or reducing its antecedent level of excitation, if the connection is of the negative or *inhibitory* variety. The absolute amount of that neuronal excitation or inhibition at any moment is a function of (a) the quasi-permanent physical size (mainly, the surface area of the contact) or *weight* of the connection at issue, and (b) the *strength* of the currently arriving signal sent up the relevant axon by the original sending neuron in the population one rung down.

Looking at this arrangement one synapse at a time reveals little or nothing of any specifically cognitive interest—rather like peering at an early-twentieth-century newspaper photo through a microscope, one boring black dot at a time. But if we draw back our focus to encompass the entire *system* of synaptic connections, and with them, the entire *population* of receiving neurons, things get more interesting in a hurry. For that population of neurons, like every other such population, is engaged in the *representation* business: any momentary pattern or distribution of excitation-levels across that collected population counts as a fleeting representation of something-or-other. Like the momentary pattern of brightness and darkness levels across the 200,000 pixels of your TV screen, it can represent a dizzying multitude of different things. Such fleeting representational patterns are simultaneously sent upward, each pattern as a whole, along the many projecting axons, to arrive at the doorstep of the next population of neurons in the ladder. At that doorstep, separating the arriving pattern from the receiving population, lie the thousands or even millions of synaptic connections, each one ready and waiting to do its assigned job of stimulating or inhibiting its own postsynaptic neuron, as a function of whatever tiny aspect of the original excitation pattern finds its way up to that tiny axonal fingertip.

If the neurons in any discrete population are engaged, collectively, in the business of representation, then the synaptic connections *onto* that population are engaged, collectively, in the business of *transforming* one representation (the one arriving from one rung down) into a *new* representation: the one embodied in the resulting excitation pattern across the receiving population of neurons. That is, a set of synaptic connections functions as a transformer, and the specific *nature* of the transformation effected is determined by the specific *configuration* of connection strengths or weights across that transforming population of synapses. If that weight configuration remains constant over time, then the synapses will collectively perform the identical transformation upon each arriving representation. Of course, since those inputs will be endlessly various, so also will be the transformed outputs. Accordingly, even if one's synaptic weights were to be frozen into some unchanging configuration, one's cognitive life might still enjoy endless novelty, simply because of the endless variety of sensory stimulation at the ladder's bottom rung.

But those synaptic connections are *not* frozen into a stable configuration of neuronal contacts and connection weights—not, at least, during the formative years of childhood. On the contrary, they are plastic and shifting. Old connections are lost and new ones made; the surface area or 'weights' of the existing connections wax and wane under the pressures of experience; and the nature of the global transformation they effect *evolves*—in pursuit, as it were, of an *optimal* transformation performed at the doorstep to each rung in the processing ladder. In short, the system learns. We may call this the Synaptic Modification Model of Learning.

Just what those assembled transformations do, and just why the process of tuning those transformations constitutes learning, are matters we shall address forthwith. We need to look at two illustrative examples, one of which involves the acquisition of the capacity for simple *categorical perception*, and the second of which displays the acquisition of a simple form of *sensorimotor coordination*.

2   Some Lessons from Artificial Neural Networks

Consider the hyper-simple feed-forward network in the center of figure 2.2. I here sacrifice several kinds of authenticity, both biological and functional, in the service of brevity and explanatory clarity. In particular, this artificial network has only two input neurons, two secondary neurons, and a single third neuron, so that we may represent their respective activation spaces with a pair of visually accessible *two*-dimensional spaces for the first two rungs, and a *one*-dimensional space for the third. Most real networks will contain rungs with many thousands of neurons, and such an activation space will thus display many thousands of dimensions; but it is difficult to draw a thousand-dimensional space, and even more difficult to portray, graphically, a specific transformation from one such space to another. So we begin with a maximally simple case, one the eye can grasp all on its own.

Training the network—so that it comes to embody some specific cognitive capacity—consists in making, in sequence, a large number of small changes to the weights of its various synaptic connections. During what is called "supervised learning," those small changes are successively imposed in response to the network's sequential performance at the third-rung or 'output' neuron, N_5. In the present case, N_5's job is to detect an instantiation of either of two *perceptual categories*—let us call them, rather fancifully, *dogs* and *cats*—as originally represented in the current activation pattern across its two perceptual neurons, N_1 and N_2. (It is fanciful because no *two* specific dimensions of perceptual variation will reliably unite all dogs, unite all cats, and distinguish both from each other and from all other animals. But never mind. Think of the two perceptual dimensions as y = size, and x = fineness-of-fur, and the pretense will not be too far off.) Our output neuron, N_5, is supposed to register a perceived instance of *dogs* by displaying a minimum level of excitation, or an instance of *cats* by a maximum level of excitation. An instance of *neither* is supposed to be

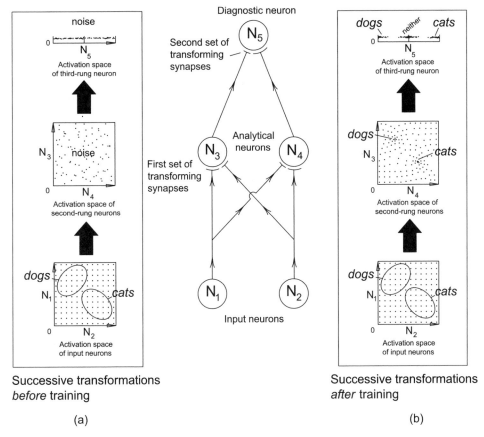

Successive transformations
before training

(a)

Successive transformations
after training

(b)

Figure 2.2
The learning of two perceptual categories by a hyper-simple feed-forward network.

signaled by an excitation level, at neuron N_5, somewhere in between those
two extremes, as when the little network encounters a squirrel, a chicken,
or a raccoon. Of course, N_5 does none of these things at first, before the
training process begins, because the network's synaptic weights have initially been set at random small values, not too far on either side from the
zero point between a positive (excitatory) value and a negative (inhibitory)
value. Whatever information may be contained in its current activation
pattern at the sensory or input rung, sending that information through a
random configuration of synaptic weights is the functional equivalent of
sending it through a mechanical paper shredder: the information disappears into a haze of noise (see the second and third activation spaces

in the transformational hierarchy to the left of the naïve network). The point of training is to improve this chaotic situation.

We can achieve that end in a variety of ways. But for illustration's sake, let us use the following procedure. Present the network, at the input rung, with an arbitrarily chosen activation pattern—a pattern, say, that genuinely falls within the range of *dog*-patterns (i.e., within the ellipsoid in the upper left portion of the $N_1 \times N_2$ activation space)—and then sit back and see what excitational behavior gets produced at the third rung, N_5. Given the random settings of the network's synaptic weights, we cannot expect appropriate dog-detection at N_5, and sure enough, it displays a resulting activation level of, say, 43 percent of maximum.

This is wrong, of course. (Recall that *dogs* are to be signaled, at neuron N_5, with something close to a 0 percent activation level.) But we (the supervising teachers) have learned *something* from this tiny experiment, namely, that the actual output (43 percent) produced by the initial weight-configuration is 43 percent above what it *should* be, if the network is ultimately to behave as we desire. Our task is now to *reduce* that 43 percent error, if only by some small amount.

This we can do by holding fixed any five of the network's six synapses, and then slightly (only *slightly*) raising or lowering the weight of the remaining student synapse, in order to see what small *improvement*, if any, either change would produce in the errant output of 43 percent, when the very same *dog*-pattern is presented once again at the sensory rung. If either fiddle reduces the initial error, that weight adjustment is retained. If neither yields any reduction in error, the original weight is left unchanged.

With this small and partial fix in place, we then move on to the next synapse, and, while holding the weights of all other connections constant, we play the same exploratory game with *that* isolated element of the transforming configuration. Repeating this process, for all six of the synapses in the network, typically results in a (very) tiny improvement in the behavior of neuron N_5—perhaps a tenth of one percent, if we are lucky. But we are only begun our training exercise. We now choose a *new* sensory input pattern—something, say, from the *cats*-region of the input rung's activation space (note the second ellipsoid toward the lower right of that space), and then repeat the suite of operations just described, this time in the attempt to reduce the *new* erroneous response at N_5. Once again, a tiny improvement is the best we can hope for at any point, but we can determine, in any event, never to make things worse, and to heed the counsel of patience. Progress delayed need not be progress denied, for there is always a new input pattern waiting in line, one that may rekindle the process of error-reduction at N_5. As training proceeds, the network is randomly shown a fair sampling of all of the possible inputs on which it may eventually be required to perform.

One pass through every input in this 'training set' of perceptual examples—each of which prompts a proprietary nudge for each synapse in the network—constitutes a single *training epoch*. Since our grid of input samples contains ($12 \times 12 =$) 144 possible activation patterns (each dot, remember, represents one pair of input values for N_1 and N_2; see again fig. 2.2), and our network has six synapses, one training epoch will involve ($6 \times 144 =$) 864 nudgings. (Although, as indicated above, some nudgings will be 'null' nudgings.) Depending on a variety of factors, successful training may involve hundreds or even many thousands of training epochs, and thus many millions of individual synaptic adjustments. But the procedure just described, known widely as the "back-propagation-of-error" algorithm, will quite regularly yield a successful weight-configuration, and highly accurate diagnostic behavior at N_5, for a wide variety of different cognitive problems. Being an algorithm, it can be automated in a computer program, and student networks like the one just described are typically taught, in fact, by an appropriately programmed and endlessly patient conventional computer.

But what is special about that hard-won configuration of synaptic weights? What makes it special beyond its ultimately producing behavioral success at the output rung? How, in short, does the system *work*?

A glance at the input-space, $N_1 \times N_2$, will reveal a grid of dots, where each dot represents a *possible* activation pattern across the first-rung neurons. Those patterns are mutually exclusive, of course (a neuronal population can display only one activation pattern at a time), but the grid as a whole represents a uniform or fair sampling of the continuous range of possible input patterns. To each one of those points in input space, there corresponds a unique point in the next space up, the $N_3 \times N_4$ activation space, a point that represents the activation pattern, in that second space, *produced by* the current configuration of synapses when they *transform* the peculiar pattern arriving from the input rung. As you can see from the $N_3 \times N_4$ space in column (a), there is no rhyme or reason to the assembled transformations performed by the randomly set weights of the untrained version of the network: they form an unprincipled scatter.

As those connection weights are progressively nudged toward an optimal configuration, however, the assembled transformations they perform slowly approximate the very different distribution of secondary activation-patterns displayed in the $N_3 \times N_4$ space of column (b). This is no unprincipled scatter. The second-rung points produced by transforming the *dog*-region points at the first rung are now clustered very close together, much closer than their 'ancestor' points in the input space. The same is true for the second-rung 'transformational offspring' of the *cat*-region points at the first rung. And a virtual no-man's-land has sprung up to *separate* those two focal clusters at the second-rung activation space.

The global transformation effected by that hard-won system of synapses yields, in column (b), an $N_3 \times N_4$ space with an *internal metric of similarity and difference relations* that contrasts sharply with the corresponding similarity metric of the $N_1 \times N_2$ input space. We can appreciate this more clearly if we magnify those successive spaces, as portrayed in figure 2.3a. You may there observe that the distance between the two circled dots in the *cat*-region of the input space is identical to the distance between the rightmost circled dot and the dot above it marked with an X. And yet, in the

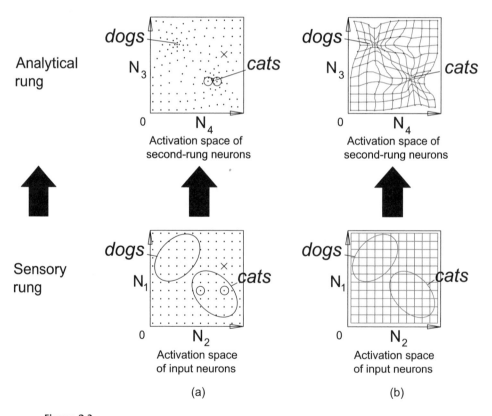

Figure 2.3
The modification of similarity metrics during learning.

next space up, their three daughter points, similarly marked, are now very differently related. The two circled points are now pressed hard against one another, and the point marked with an X is now a comparatively long way from either. That is to say, the former two are represented, at the upper space, as being highly *similar* to one another, while the latter (the X) is represented as being very *different* from both. What were equal distances in the input space have now become very different distances in the next space up.

This gives us, incidentally, a plausible explanation for so-called 'category effects' in perceptual judgments. This is the tendency of normal humans, and of creatures generally, to make similarity judgments that group any two within-category items as being much more similar (to each other) than any other two items, one of which is inside and one of which is outside the familiar category. Humans display this tilt even when, by any "objective measure" over the unprocessed sensory inputs, the similarity measures across the two pairs are the same.

The artificial networks recreate this important cognitive profile partly because their artificial 'neurons,' like real neurons, typically embody a nonlinear squashing function in their activational responses to arriving synaptic influences. Specifically, in the middle range of its possible excitation levels—roughly between 25 percent and 75 percent—a neuron is happy to have its activation modulated up or down by an amount that is directly proportional to the sum of the several synaptic influences arriving at its doorstep. But as one approaches the two extremal points of possible activation—0 percent and 100 percent—the neuron displays an increasing resistance to synaptic modulation (fig. 2.4). Figuratively speaking, you have to push like mad to get the neuron off the ground and flying in the first place; gaining altitude thereafter is easy for a stretch; and then a ceiling is approached where even large expenditures of synaptic stimulation produce only small increases in altitude.

This otherwise boring wrinkle in the response profile of neurons yields a profound benefit. If that profile were always a straight line instead of the sigmoid curve of figure 2.4, the all-important transformations effected by a cadre of synapses would be limited to what are called *linear* transformations. These will effect changes in the grid of the input space that amount to uniform rotations, translations, or trapezoidal tiltings of that initial grid, but they are incapable of effecting *complex curvilinear* changes. The curvilinear character of the sigmoid profile lifts the entire transformational system of a network into the much larger realm of strongly nonlinear transformations. Equally important, the variously overlapping outputs of

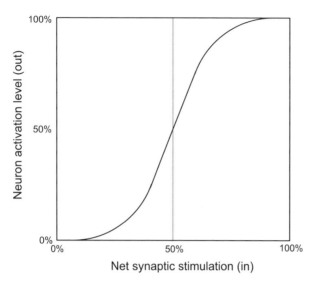

Figure 2.4
The 'squashing function' for modulating neuronal output activation levels.

hundreds and thousands of differently activated nonlinear neurons allows
for the accurate *approximation* of almost any nonlinear transformation that
the network might need to perform.

The reader may observe directly an instance of such a nonlinear transformation in the little cartoon network already on the table. Figure 2.3b
illustrates how the rectilinear grid of its original input space gets transformed into the variously squashed and metrically distorted activation
space at the second rung. You can now see *why* the dots of figure 2.3a get
redistributed, into two prominent clusters, in the transformation from the
first to the second space: it is because the metric of the input space *as a
whole* has been warped or transformed into the decidedly nonlinear space
of possibilities at the next layer up. As you can see, the resulting $N_3 \times N_4$
space embodies two narrow "hot spots" or "prototype regions" where the
daughter paths of the original input geodesic paths are now pinched very
close together. The job of our final or "diagnostic" output neuron, N_5, is
now the relatively simple one of producing a minimal response to anything in the close neighborhood of the *dogs* hot-spot in the second activation space, a maximal response to anything in the *cats* hot-spot, and a
ho-hum intermediate response to anything else.

In the $N_3 \times N_4$ space of figure 2.3b, we are looking, at last, at the
conceptual framework that training has produced in our cartoon network.

That space embodies the framework of *categories* into which the mature network divides its perceptual world, a metric of *similarity* and *difference* relations between any two perceptual inputs, and a presumptive metric of *probabilities* across that space of activational possibilities. The regions of the two prototypical hot spots represent the highest probability of activation; the region of stretched lines in between them represents a very low probability; and the empty regions outside the deformed grid are not acknowledged as possibilities at all—*no* activity at the input layer will produce a second-rung activation pattern in those now-forsaken regions.

3 Motor Coordination

Adding nonlinear transformations to the repertoire of rung-to-rung computations yields payoffs not just at the perceptual end of the system, but at the *motor* end as well. Consider an example I have used before, and have recently found reason to use again.[1] The fictive crab of figure 2.5a (plate 3) has two eyes, which rotate on a vertical axis only. Jointly, they can register the objective physical position of any perceived object, in the horizontal plane in front of it, by the pair of eye angles assumed when the object is jointly foveated, as indicated. The crab also has a moveable arm with two joints: an "elbow" and a "shoulder." The momentary configuration of the arm in objective space is uniquely specified by the two angles jointly assumed at that time, as indicated in figure 2.5b.

What the crab now needs, to make use of this equipment, is some way to convert or transform information about the spatial location of an edible

1. The reason is the discovery, by Graziano et al., of a map—a map of the body's many possible limb-positions—embodied in the motor cortex of the macaque. Specifically, the simultaneous electrical stimulation of the local motor neurons surrounding a given point on the surface of the monkey's primary motor cortex causes the monkey's limbs to assume a configuration specific to that 'geographic' position on the cortical surface, regardless of the body's limb configuration prior to the stimulus. Stimulations at proximate points yield closely similar limb configurations, and stimulations at distal points produce quite disparate limb configurations. In all, the cortex embodies a map of the animal's 'limb-configuration space.' (A caution. We should not assume, from the two-dimensional nature of the cortical surface, that its embodied map must also be two-dimensional. Each 'pointlike' electrical stimulation activates, to varying degrees, a nontrivial *population* of distinct motor neurons surrounding that point, and it is likely the *profile* of activation levels across a large number of neurons that codes for each limb position. The map, in short, is likely a high-dimensional map.) See Graziano, Taylor, and Moore 2002.

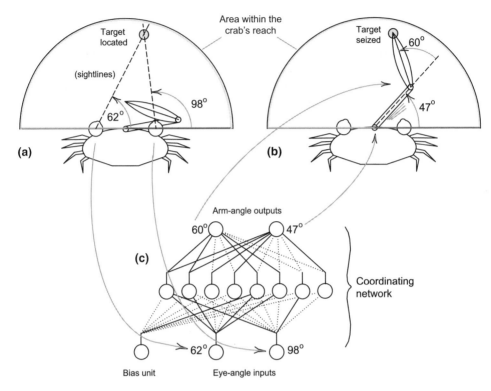

Figure 2.5
Sensorimotor coordination achieved by a map-transforming neuronal network. See plate 3.

tidbit, information whose initial format is a pair of tidbit-foveating eye-rotation angles, into the quite different format of a unique pair of arm angles, a pair that will position the arm so that the tip of the forearm touches exactly the spot where the target tidbit awaits. He needs, in short, some sensorimotor coordination.

Enter the network of figure 2.5c. It has two neurons at the sensory input rung, one for each eye, each of whose current activation level represents the current rotation angle of its corresponding eye. The network also has two output neurons, one for each joint, whose activation levels are supposed to dictate a corresponding pair of joint angles in the crab's moveable arm. They have, let us suppose, direct causal control over those two angles (that is, let us suppose that they are *motor* neurons). Between the input and output rungs there is a middle rung of eight neurons with the appropriately nonlinear sigmoid response profiles discussed earlier, neurons whose collective job it is to try to approximate the strongly nonlinear transformation required to put the crab in business.

Part of the point of making this crab so simple is that two of its important activation spaces—the input space and the output space—are deliberately two-dimensional, and thus we are again in position to grasp, with the naked eye, how things are represented therein, and how the original information in the $N_1 \times N_2$ eye-angle sensory-input space gets transformed in its ultimate re-presentation at the $N_{11} \times N_{12}$ arm-angle motor-output space. Figure 2.6a (plate 4) displays the orthogonal grid of the sensory input space, with the upper-left portion in a grid of solid lines to indicate the region of activation pairs that correspond to positions in objective space *within reach* of the crab's two-jointed arm (the D-shaped region in fig. 2.5).

The dotted-line region corresponds to pairs of eye-directions that do not intersect at all ("wall-eyed" positions), or to eye-directions that intersect all right, but beyond the maximal reach of the crab's outstretched arm. Note well the red, blue, and yellow borders, which correspond to the same-color spatial regions in the original figure 2.5a and b. Note also the bold-faced red and green rectangle regions, just so that you may reidentify them, transformed but still boldfaced, within the motor output space.

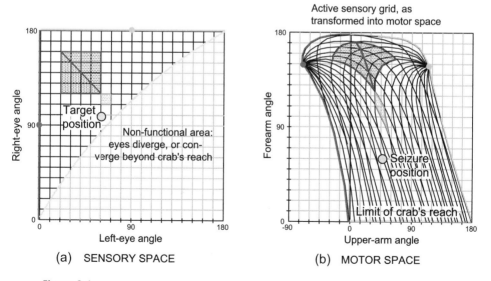

Figure 2.6
The metrical deformation of active sensory space into active motor space. See plate 4.

Figure 2.6b portrays the motor output space of the trained network, with one dimension for each of the arm's two joint-angles. The curvilinear grid of lines within that space represents the original solid-line visual input-space grid of "reachable positions" *after* its transformation at the hands of the two intervening layers of synaptic connections. As you can see, the two orthogonal axes of that original input space have here been shrunk to a pair of radiating points, and the two color-coded squares, now severely squashed, will help you reidentify the rest of the original space. That global transformation has the following vital property (vital, anyway, for the hungry crab). Given *any* point within the solid-lined portion of the input space (each of which represents, recall, a unique location in the very real space in front of the crab), its *daughter* point within the motor output space represents *the unique arm position that places the tip of the crab's pincer in contact with the object at that perceived spatial position.* That activation-pair at the output rung, which is a population of *motor* neurons, then causes the arm to assume exactly the pair of joint angles that they respectively represent. And thus the arm assumes a position that allows the crab to seize the perceived object.

As with the cartoon network of figure 2.2, the crab's coordinating trans-formation resides in the collective effects of its thirty-two intervening syn-aptic connections, whose precise configuration of weights was achieved by the error-reducing procedure—the back-propagation algorithm—discussed earlier. But the crab's network is not just an illustrative cartoon. Though fictive rather than real, the crab's visual and motor systems were modeled in a digital computer, as was the artificial neural network that coordinates them, and the whole system was (very) slowly taught, by successive correc-tions after successive examples, to embody the visuomotor skill at issue.[2] Unlike my opening cartoon network for dogs and cats, this network really exists, if only as a model within a digital computer, and it can direct an on-screen graphical crab reliably (if rather clumsily) to reach out and touch an object presented to it anywhere in its proximate visual world.

2. This dates to 1983, and the machine originally involved had a Zilog-80 CPU and 64K of RAM (a NorthStar *Advantage*, from an agreeable California company eclipsed long ago by the current giants). The crab's *initial* 'brain' was an explicitly geometrical transformer called a "state-space sandwich," but, following a sympathetic suggestion from David Rumelhart, I subsequently replaced it with the feed-forward artificial neural network shown here. By then, I had acquired an Intel 80286 machine. With these still paleolithic resources, the training of this small network (in 1986) took 72 laborious hours of computer time. Current desktop machines would do it in under twenty minutes.

The cognitive achievement here portrayed—a particular form of visuo-motor coordination—is quite evidently an acquired *skill*. And on that account, there may seem little that is distinctly *cognitive* about it, in the sense that is likely to interest an epistemologist. But note that the cognitive achievement of the perceptual network discussed earlier—the ability to recognize and report instances of two perceptual categories, *dogs* and *cats*—is also no less an acquired skill, and the exercise of one's perceptual skills is a quintessentially epistemological matter. Moreover, Pragmatists such as C. S. Peirce have been telling us for years that an acquired *belief* is, at bottom, nothing but an acquired *habit of action*.

Whether or not such a simple reductive analysis might be made to stick (I doubt it), its proposal signals the poverty of any sharp or fundamental distinction between knowledge *how* and knowledge *that*. There is a grammatical point to the distinction, to be sure: the latter relative pronoun is properly completed by a declarative sentence, while the former is properly completed by an infinitive verb phrase. But whether this superficial grammatical distinction corresponds to any important or fundamental difference in representational formats within the brain remains to be seen. The suggestion to be pursued in this essay is that the major fault lines that divide our neurobiological styles of representation lie quite elsewhere, and largely cross-classify those embodied in Folk Psychology. In particular, the brain's representations of the world's enduring categorical and causal structure (its "factual" knowledge), and the brain's representations of its various acquired motor skills and abilities (its "practical" knowledge), are *both* embodied in the carefully sculpted metrics of similarities and differences that provide lasting structure to each one of the brain's many activation spaces.[3]

3. Indeed, in some cases one and the same activation space appears to do 'double duty' by serving as the canvas of representation for both motor-behaviors-as-perceived-in-others, and for motor-behaviors-as-produced-by-oneself. I have in mind here the population of so-called 'mirror neurons' observed in macaques by Rizzolatti, Fogassi, and Gallese (2001). Single-cell recordings from such neurons, in the prefrontal cortex of awake, behaving animals, display identical activational behaviors whether the animal is voluntarily generating a certain behavior itself—such as picking up a nut from an outheld palm—or whether it is observing the same behavior as performed by another animal. (In the latter case, of course, there must be some inhibitory activity somewhere in the downstream cognitive system, lest the creature compulsively mimic every motor behavior it sees.) It is just possible, of course, that said population of neurons has a *purely* perceptual function in both cases. But that frontal region is known to be a motor-assembly area (that's why Rizzolatti's lab was studying it), and lesions to it disrupt motor control at least as much as they disrupt motor perception.

Correlatively, the brain's representations of the perceptual world's current and specific configuration here-and-now, and the brain's representations of the body's current and specific motor undertakings here-and-now, are *both* embodied in the fleeting *activation-patterns*, here-and-now, that occur at some specific point *within* those background activation spaces. The principal epistemological distinction confronting us here lies between one's lasting *background* knowledge on the one hand, and one's current and *ephemeral* representations on the other. The former category encompasses one's general knowledge of both the world's enduring structure and how to navigate it, while the latter encompasses one's current perceptual and one's current volitional activities.

Interestingly, both one's motor skills and one's perceptual skills are typically inarticulate, in that one is unable to say, for example, *how* one recognizes a color, or the gender of a face, or the sound of a flute; likewise, one is unable to say *how* one manages to utter a simple sentence, or catch a bean bag, or whistle a tune. As we say, one just can. The reason for this inarticulation is that, in both cases, the relevant wisdom is embodied in the configuration of thousands or millions of synaptic weights, and thereby in the complex metrical structure of the activation space to which they connect. But neither of these intricate circumstances is typically accessible to the person whose brain is their home.

Philosophers such as Bert Dreyfus, John Searle, Fred Olafson, and Avrum Stroll have often spoken about the unarticulated but ever-present "Background"—that is, our background grasp of the world's prototypical structures, and our background grasp of how to navigate our way around and among those structures. We are here looking at the neurocomputational *basis* of that Background. Its basis is the overall configuration of synaptic weights that learning has produced in the brain, and it is constituted by a complex, ladder-like hierarchy of artfully sculpted neuronal activation spaces. That hierarchy embodies the multilayered conceptual framework of each and every one of us—the background representational and computational framework within which, after reaching maturity, one's every cognitive move is fated to be made.

4   More on Colors: Constancy and Compression

An early portrait of the space of possible human color-experiences pictured in figure 1.3 (plate 1) was pieced together over a century ago by a psychologist named Munsell, and by a process that had nothing to do with tracing the activational behavior of neurons. Munsell simply showed a large number of three-element color samples to his experimental subjects, and asked them to specify which two of those color samples were more similar to each other than to the third sample in that triplet. Not surprisingly, different subjects showed closely similar judgments on this task, across the entire range of samples. He subsequently noted that the *global* pattern of similarity-and-difference relations thus uncovered could be simply represented by a set of *distances* in a unique three-dimensional solid with white at the top, black at the bottom, and the familiar handful of maximally saturated colors ordered in a rough circle around the center of the white-to-black vertical axis. All other colors are located at some point with a unique family of distance relations to those several landmark colors. Munsell, too, noticed the substantial tilt in that "equator" (see again fig. 1.3, plate 1), a tilt that places saturated yellow higher up in the solid, much closer to white than to black, and which also places saturated blue and purple lower down, much closer to black than to white. Overall, the similarity or dissimilarity of any two colors was simply represented by their mutual proximity or distality within that space. This spindle-like solid with its wonky equator is standardly called the *phenomenological color space* for humans.

The *explanation* for the curious structure of this space had to wait some eighty years before finally emerging, in the form of a simple two-runged neuronal processing ladder proposed by L. M. Hurvich and D. Jameson (see fig. 2.7a, plate 5). By then, the existence of three distinct populations of wavelength-sensitive neurons scattered across the retina—the three kinds of rather particular *cone*-cells, as distinct from the wavelength-indifferent *rod*-cells—had become common knowledge. These were maximally sensitive to light at 0.45 μm (short wavelength, seen as vivid *blue*), 0.53 μm (medium wavelength, seen as a yellowish *green*), and 0.58 μm (long wavelength, seen as an orangey *red*), respectively. Collectively, they form the input population or first rung of the human color-processing system.

But *only* the input. As you can see in figure 2.7b (plate 5), the response profiles of all three kinds of cones are very broadly tuned: they have a maximum activational response at the three wavelengths indicated, but they will still respond—though progressively less enthusiastically—to light at wavelengths on either side of that 'preferred' position. Note also that their respective response profiles overlap one another, substantially for the medium (M) and long (L) cones, and slightly but still significantly for the M and the short (S) cones. Collectively, they are sampling the *distribution* of energy across *all* wavelengths in the visible domain, and the results of that sampling are sent upward to the next population of neurons, for

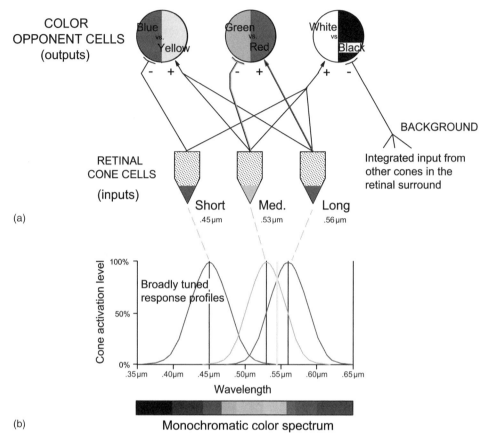

COLOR
OPPONENT CELLS
(outputs)

RETINAL
CONE CELLS
(inputs)

(a)

BACKGROUND

Integrated input from
other cones in the
retinal surround

(b)

Figure 2.7
The human color-processing network (after Jameson and Hurvich 1972). See plate 5.

processing and analysis by the cadre of synaptic connections at its doorstep.

The configuration of synaptic connections in the schematic net proposed by Hurvich is unusually uniform and simple: half are excitatory, half
are inhibitory, and they all have the same absolute weight. What is interesting is how they place the three input neurons into a variety of tug-of-
wars for control of each of the three classes of neurons at the second rung.
Consider, for example, the middle neuron at the second rung, the so-called
'red-vs.-green' cell. Its only inputs are an excitatory projection from the
L-cone, and an inhibitory projection from the M-cone. In the absence of
any input from these two sources, the receiving cell has a default or resting
activation level of 50 percent. Any level above or below that median is

therefore a direct measure of the relative preponderance of incident light in the M-region over (or under) incident light in the L-region, at the tiny area on the retina where the three input cones are clustered together.

Similarly, the left-most neuron, the so-called 'blue-vs.-yellow' cell, is the site of a tug-of-war between an inhibitory projection from the S-cone, and an excitatory projection jointly from the M- and L-cones. Here, any activation level above or below its resting level of 50 percent is a barometer of the preponderance of incident light in the M- and L-regions over (or under) incident light in the S-region—once again, all at the same tiny area on the retina.

Finally, the right-most or "black-vs.-white" cell receives an excitatory projection from *all three* of the cones at that point on the retina, and a "background" inhibitory projection from some mechanism that averages the total amount of light, at any and all wavelengths, over a rather larger retinal area *surrounding* the retinal point at issue. Its current activation level thus reflects the current *contrast* between the local brightness-level at the site of the three cones at issue, and the average brightness across its background surround. These three kinds of cells—the RG (red–green), the BY (blue–yellow), and the BW (black–white)—are called, for obvious reasons, "opponent-process" cells, or "color-opponent" cells.

We may summarize this arrangement with the following three linear equations, expressing the activation level of each of the three second-rung opponent-process cells as a function of the several activation levels of the first-rung cone cells that project an axon to them. ("B" here represents the *background* brightness—the *sum* of the average levels of L, M, and S activation—across the larger area surrounding the three cone cells at issue.) The numbers contained in these equations reflect (1) the fact that each of the three second-rung cells has a resting activation level of 50 percent of maximum; (2) the requirement that the activation levels of each of those cells range between 0 percent and 100 percent; (3) the requirement that the polarities of the several synapses are indicated in figure 2.7; and (4) the requirement that each of the three tug-of-wars portrayed above be an even contest.

$$A_{RG} = 50 + (L - M) / 2$$

$$A_{BY} = 50 + ((L + M) / 4) - (S / 2)$$

$$A_{BW} = 50 + ((L + M + S) / 6) - (B / 6)$$

If we plot these three equations in a three-space, the only points (i.e., the only activation patterns across those three cells) possible are those within the trapezoid-faced solid shown in figure 2.8a. The cut-gem character of this solid reflects the fact that we used three *linear* equations as our first-pass approximation for the activational behavior of the second-rung cells. A more realistic model would multiply each of A_{RG}, A_{BY}, and A_{BW} by a sigmoid squashing function, as discussed above. That modest wrinkle has the effect of rounding off some of the sharper corners of the linear model, which yields the more realistic solid shown in figure 2.8b (plate 6). (I here suppress the relevant algebra.)

To continue, let us make a deliberate cut through its 'equator,' that is, a planar cut within that solid, as indicated in figure 2.8c (plate 6). That internal plane is now explicitly painted so as to indicate the various external colors actually coded by the various points (i.e., the three-element activation patterns) within that central plane. You will recognize here a strikingly faithful reconstruction of the original phenomenological color space, as pieced together by Munsell. It even reproduces the strongly tilted equator, with maximally brilliant yellow being coded at the upper right, and maximally saturated blue coded at the lower left. That solid is evidently a *map* of the space of all possible objective colors.[4]

But what is the functional *point* of transforming, recoding, and compressing the original retinal-cell information from four retinal dimensions to three opponent-cell dimensions? What cognitive advantage does it buy the creature that possesses such a processing network? There are several, but perhaps the first is the gift of *color constancy*, the capacity to see an object's objective color, for what it is, across a wide range of levels of background illumination. From bright sunlight, to gray overcast, to rainforest gloom, to a candle-lit room, a humble green object will look plainly and persistently green despite the wide differences in energy levels (across those four conditions of illumination) reaching the three cone cells on the retina.

This stabilizing trick is turned by the "differencing" strategy used by the Hurvich network. Thanks to the tug-of-war arrangement discussed above, the network cares less about the absolute *levels* of current cone-cell activation than it does about the positive and negative *differences* between them. An input pattern of $\langle L, M, S \rangle = \langle 5, 40, 50 \rangle$ will have exactly the same effect as an input pattern of $\langle 15, 50, 60 \rangle$, or $\langle 25, 60, 70 \rangle$, or $\langle 43, 78, 88 \rangle$,

---

4. It is also a map, if a somewhat low-resolution map, of the range of *possible electromagnetic reflectance profiles* displayed by physical objects. This is one of the reasons motivating the reductive identification of objective colors, the reflective ones anyway, with EM reflectance profiles in the window 0.40 mm to 0.70 mm. For the details, and a proposed solution to the problem of color metamers, see ch. 10 of Churchland 2007a.

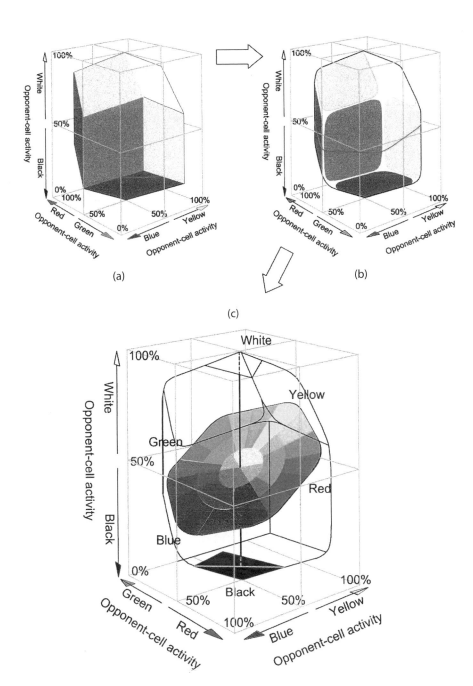

Figure 2.8

The solid neuronal activation space of the Hurvich-Jameson network, and its internal map of the colors. See plate 6.

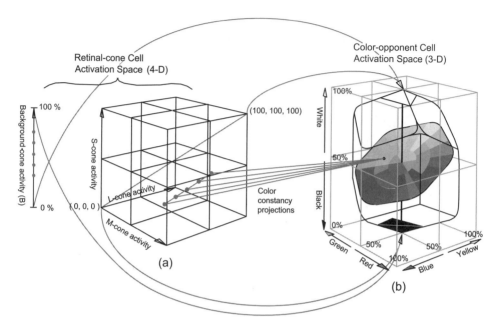

Figure 2.9
Color-constancy in the Hurvich-Jameson network. See plate 7.

namely, a stable output pattern, at the second layer, of $\langle A_{BY} = 36.25$, $A_{RG} = 32.5$, $A_{BW} = 50\rangle$, specifically, a grayish-green. At least, it will do so if the increasing activation values cited are the result of a corresponding increase in the level of *general* background illumination, because in that case the absolute value of B (which codes for background brightness) will *also* climb, in concert, by exactly 10, 20, and then 38 percentage points as well. (See esp. the last of the three equations above.) A color sensation that might otherwise have climbed steadily, into the region of a pale pastel chartreuse toward the upper apex of the color spindle, thus remains nicely fixed at the true color of the external object, a dull green. (See fig. 2.9, plate 7.) And so for all of the other colors in the spindle.

We can now begin to appreciate the global nature of the synapse-induced transformation from the input retinal-cone space to the opponent-cell space, which latter is the color space proper. Not only is the geometrical origin of the retinal-cone space re-represented at the center of the color-space's basement floor, and the retinal-cone space's maximum activation triplet re-represented at the center of the color-space's ceiling, but what are so many *extended straight lines* within the input retinal-cone space get shrunk to so many *single points* within the second-rung color space proper. This latter is what we saw happening in the preceding paragraph.

This clever system allows us to see past the ephemeral vagaries of raw sensory stimulation to grasp an enduring objective reality. This is a theme that runs through our perceptual mechanisms generally, and we can here point to the first advantage of the information-compression effected by the Hurvich network: it gives us grip on any object's objective color that is independent of the current background level of illumination.

Putting color aside for a moment, let us note that a visually equipped creature with no color perception at all would still confront the same basic problems as the rest of us. Specifically, a color-blind creature needs to get a fix on the objective spatial positions, boundaries, shapes, and motions of local physical objects *in the teeth of* the following confounding factors: (1) the variations in the levels of illumination across the objects (due to their variously curved and oriented surfaces, and to the scattered and often moving shadows on them, etc.), and (2) the complex and changing partial occlusions of many objects by many others. Finding order in this chaos is the first job of any visual system.

Since most unitary objects in the world are not deliberately camouflaged, an aspiring or upwardly mobile visual system has a promising strategy to pursue. That is to say, since most objects—like a rock, a tree trunk, a leaf, an animal, a body of water—have a disposition to reflect incident light that is roughly *uniform* across its surface, the visual system must learn to see *past* the vagaries cited in the previous paragraph. It must learn to identify surfaces of *constant* objective reflectivity, and thereby begin the business of segmenting the chaos of one's visual field into distinct—if variously lit, curved, and oriented—physical objects.[5] The visual systems of color-blind creatures—which perform a purely 'gray-scale' analysis—are already exquisitely skilled at doing precisely this.

But those color-blind creatures confront potential confusions and ambiguities when distinct objects have *identical* objective reflectance efficiencies (for the extreme case, imagine an environment in which absolutely everything is the same objective shade of middle-gray). Vision is still entirely possible under such circumstances, but borders and boundaries that were distinguishable *only* by their grayscale reflectance differences will now be mostly invisible. Color vision, by contrast, can avoid such confusions and

5. Artificial networks have succeeded in this cognitive task as well. I commend to the reader's attention the "Shape from Shading" network described by Sejnowski and Lehky (1980).

defeat such invisibilities, because two such objects are likely to *differ* even so in the wavelength *profile* of their reflectance efficiencies, even if they are the same in the total fraction of incident light energy that they reflect, that is, even if they are grayscale equivalent. Equally important, most unitary objects—such as leaves, tree trunks, and animals—have a wavelength reflectance profile (i.e., a color) that is roughly *uniform* across their outer surface. This allows the trichromatic visual system of humans to get an *additional* grip on objective surface uniformities and differences, and thus to do a better job of the visual system's original and still central function—segmenting correctly the world of physical objects.

On this view, humans and other animals did not develop a trichromatic visual system "in order to perceive the colors of things," but rather to further enhance the efficiency of the more basic tasks of object segmentation and movement detection. It does this by giving us a sensitivity to *samenesses* and *differences* in the proprietary *wavelength profile* of the objective reflectivity of distinct physical objects, and therefore a better grip on the objects themselves. I owe this insightful take on the basic function of color vision to Kathleen Akins (see Akins 2001). It gives any visual system a second reason, beyond an indifference to changing light-levels, to welcome the peculiar transformational character of the Hurvich network.

There is a third reason to welcome it. If one is limited, as we are, to only three distinct kinds of opponent-process cells, then the roughly double-coned shape and internal structure of the color space produced by the Hurvich arrangement may be (or is at least very close to being) *the most efficient possible way* of representing, in compressed form, information about the wavelength reflectance-profiles of visible objects, or at least, the profiles characteristic of the visible objects most commonly found in our normal terrestrial environment. This conclusion is strongly suggested by an experimental result from a related network-modeling effort by Usui, Nakauchi, and Nakano (1992). Usui and his coworkers constructed a 'wineglass' feed-forward network whose 81 input 'cone cells' sampled an input reflectance profile at 81 distinct wavelengths (see fig. 2.10). Its 81 output cells—five rungs away from the input cells—were supposed to simply reconstruct/repeat the current pattern of stimulations entered as input. (This is one instance of what is called an *auto-associative* network, because the input and output patterns in the successfully trained network are required to be identical. Such networks are used to explore various techniques of selective *information compression*.)

This task might seem simple enough, but the network was deliberately constructed so that all of the input information was forced to go through

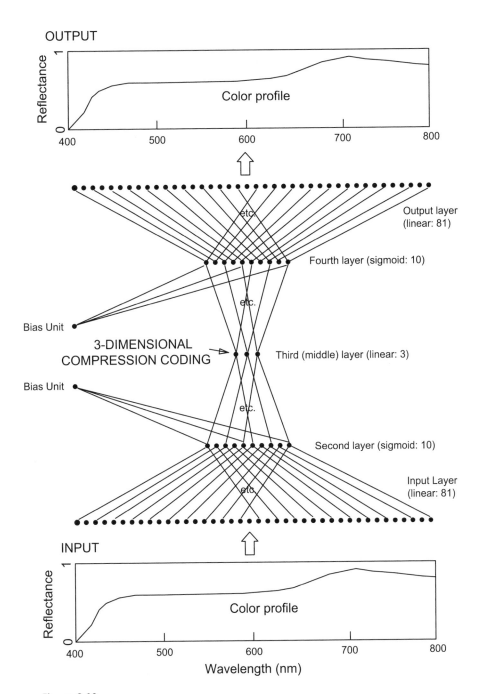

Figure 2.10
The 3-D compression of a complex wavelength profile in a color-coding network
(after Usui, Nakauchi, and Nakano 1992).

a bottleneck of only *three* cells, before being successfully reconstructed at the output rung. This might seem to be flatly impossible, and indeed, much information is inevitably lost. But after supervised training, by the back-propagation procedure, on a large number of color samples (more specifically, on the 81-element *sampling* of the wavelength reflectance-profiles that they display), the network did indeed learn to reproduce, at the output rung, highly creditable copies of the many input profiles on which it was trained, and of novel color inputs as well.

But how does it manage to do this? In particular, what is now going on at the three-celled middle rung? Into what clever coding strategy did it gradually relax during the long training process? Fortunately, since this is an artificial network, its activities can be probed at will. Addressing the now-mature network, Usui and company simply tracked the sundry activation triplets produced at the middle rung by a wide variety of input colors, and then plotted the input colors variously coded by those triplets in a three-space, a space with one axis for each of the three middle-rung cells (see fig. 2.11).

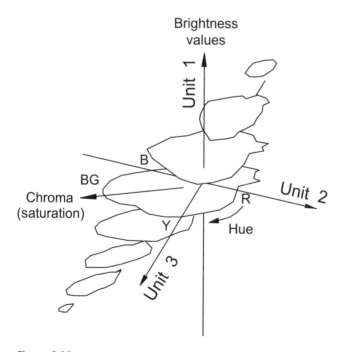

Figure 2.11
The color-coding space at the middle layer of the Usui network (after Usui, Nakauchi, and Nakano 1992).

The activation triplets across the critical middle rung were all confined to a roughly spindle-shaped solid, with a black-through-gray-to-white central axis, and with the saturated versions of the familiar colors arrayed in sequence around its tilted equator. The artificial network, trained on the same color inputs encountered by humans, fell into roughly the same coding strategy employed by human opponent process cells. And it did this, note well, despite having *81* distinct and narrowly tuned types of "cone cells" at the input rung, where humans have only *three* distinct and rather broadly tuned types of cone cells within the retina: the L, M, and S cells. But the relevant higher-level rungs, in the artificial network and in us, learn/employ roughly the same coding strategy even so.

That coding strategy is far from perfect: there are a great many distinct possible reflectance profiles between which it is unable to distinguish. (Color scientists call those same-seeming profiles "metamers.") That is no surprise. Information-compression networks succeed, when they succeed, because they forsake the impossible task of trying to compress *all possible* input information. Instead, they focus their sparse resources on compressing only the special range of inputs that they typically encounter or are somehow asked to deal with. Fortunately, our terrestrial environment does not display to us every possible reflectance profile, but only a comparatively small and recurring subset of them. And those, we *can* get a systematic and objective grip on: that is precisely what the three-dimensional analysis, embodied in the color solid, does for us. Moreover, the fact that both the artificial network and all normal humans settle into this same coding strategy suggests that it is at least a local optimum, and perhaps even a global optimum, in the space of possible solutions to the coding problem we all face.

Here, then, are several functional virtues of the peculiar transformational trick turned by the Hurvich–Jameson opponent-process network: color constancy over different levels of illumination, improved object segmentation, and optimum or near-optimum coding efficiency, relative to the resources at hand. The account at issue also provides a wide range of novel experimental predictions about the details of color vision and color experience, some of them quite surprising. (See Churchland 2007, ch. 9, for an exploration of an unexpected and anomalous class of impossibly colored after images, as they are predicted by the Hurvich–Jameson network.) But now we need to move on. In particular, we need to look again at a network with a much higher-dimensional activation space, a compression network once again, a network that embodies a conceptual framework with a hierarchical structure. This is Cottrell's face-discrimination network, briefly scouted in the first chapter.

5 More on Faces: Vector Completion, Abduction, and the Capacity for 'Globally Sensitive Inference'

Cottrell's face-discrimination network has 4,096 grayscale sensitive 'retinal' cells, arranged in a 64 × 64-element grid, on which various images can be 'projected.' Each of those many retinal cells projects an axon upward to a second rung of 80 cells, where it branches to make 80 distinct and variously weighted synaptic connections, one for each of those waiting cells. In the initial incarnation of the network, that second rung projected finally to a third rung identical in structure to the input population (see fig. 2.12a). We are looking once more at an auto-associative "wine-glass" network, like the Usui color network. This one is simpler in having only three rungs instead of five, but it is also decidedly more complex in having a middle or 'compression' layer of 80 cells instead of only three. This reflects the more complex domain it must learn to represent: faces—male faces, female faces, young faces, older faces, Asian Faces, African faces, European faces, smiling faces, grim faces, and more faces. But only faces. It is not possible

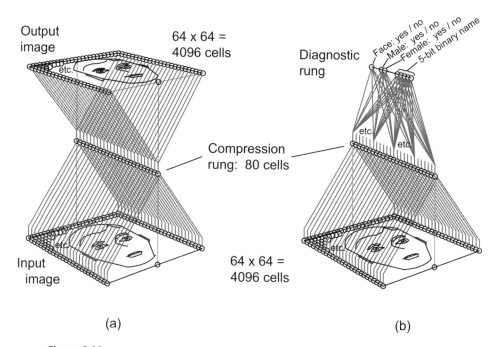

(a) (b)

Figure 2.12
(a) A network for compressing face-images into an 80-D space, and then reconstructing it. (b) A network for recognizing faces.

for the network to find a coding strategy at the middle rung that will compress any arbitrary input and then successfully reconstruct it, *from* this compressed 80-element representation, at the third or output rung. A random-dot input pattern, for example, will utterly defeat it, since any truly random sequence of values knows no expression, of any kind, more compact than that random sequence itself. But if the network's reconstructive task is confined to human faces only, the range of possibilities, though still large, is dramatically reduced. For example, all faces are very closely symmetric around a vertical axis between the eyes, and so the network's coding strategy might safely ignore one-half of the input pattern: it is, in that way, substantially redundant. As well, all faces have one nose, one mouth, two eyes, two nostrils, an upper and lower lip, and so forth. Accordingly, the mature network need not prepare itself anew, in each case, for any representational demands in conflict with these regularities. It can focus its coding efforts on the subtle *variations*, on those enduring themes, that each new face presents.

Indeed, finding a successful format for the *compression* of information about some frequently encountered domain is equivalent to, it is the same thing as, finding the enduring themes, structures, or regularities characteristic *of* that domain. We may think of the network's achievement, during the training phase, as the construction of an *abstract map* of the important samenesses and differences between the typical elements slowly discovered within the perceptual domain at issue. Once that background map is in place, the sensory-induced activation of a specific *point* within that map constitutes the network's perceptual knowledge of *which* of the many acknowledged possibilities, within the now-familiar domain, it is currently confronting.

But let us focus, once more, on the background map itself. Its hierarchical structure was crudely portrayed back in figure 1.2, and its acquired partitions were briefly discussed on that earlier occasion (pp. 8–9). But how is this precious representational structure achieved? To repeat the question asked of the simpler networks examined above, what sort of transformation is being effected, by the cadre of $(4096 \times 80 =) 327,680$ synaptic connections meeting the middle rung, so as to yield the peculiar 'facial map' at issue? Here no simple diagram, like those of figures 2.3 and 2.9, can adequately convey what is going on: the dimensionality of the spaces involved (4,096 compressed into 80) is now far too high.

Even so, we can begin to get a feel for some of what is happening if we ask: what is the *preferred input stimulus* for each one of the 80 cells at the second rung? The preferred stimulus, of a given second-rung cell, is defined

Figure 2.13
The 'preferred stimuli' for an arbitrary six of the 80 face-compression cells.

as the specific *pattern of activations* across the input or "retinal" population
that produces, via that cell's proprietary and peculiar 4,096 synaptic connections, the *maximum* level of excitation at that second-rung cell. It is the
input pattern that makes that cell 'yell its loudest.' As it happens, that
preferred stimulus, for any given cell, can be directly reconstructed from
the learned configuration of the 4,096 synaptic connections *to* that cell.
And since this is an artificial network, modeled in every detail within a
digital computer, that information is directly available, for each and every
cell, at the punch of a "status report" command.

The preferred stimuli for an arbitrary six of the 80 compression-rung
cells, as subsequently reconstructed by Cottrell, are displayed in figure
2.13. Perhaps the first thing to notice about each 64 × 64–element pattern
is that it is quite complex and typically involves the entire array. No second-rung cell is concerned, for example, with only the mouth, or only the
ears, or only the eyes. *Whole faces* are the evident concern of every cell at
this compression layer. Second, all of these preferred patterns, or almost
all, are decidedly facelike in a variety of vague and elusive ways. And that
is the key to how the network gets a useful grip any particular face-image
entered as an input pattern—your own face, for example. The network
does an 80-dimensional assay of your peculiar facial structure, as follows.

If your face image corresponds very closely to the preferred stimulus-pattern or 'facial template' of second-rung cell 1, then cell 1 will be stimulated thereby to a high level of activation. If your face corresponds only very poorly to that preferred pattern, then cell 1 will respond with a very low activation level. And so for every other cell at the second rung, each of which has a distinct preferred stimulus.

The result, for your unique face, is a unique pattern of activation levels across the entire second-rung population, a pattern that reflects the outcome of the 80 distinct similarity-assays performed in response to your face. Your face has earned a proprietary position or point within the 80-dimensional activation space of the second-rung neuronal population. But that peculiar position is not primarily what concerns us here. What concerns us is the fact that any face that is *similar to* your face will receive a *similar* assay by the 80 diagnostic cells at issue, and will be coded, within their activation space, by a position or point that is geometrically *very close to* the point that codes your face. Alternatively, a face very unlike your own will provoke a very different set of individual similarity-judgments (i.e., template matches, or mismatches) at the second rung, and will be coded by an activation point that is geometrically quite *distant from* the point that codes your own. In the mature network, the second-rung space as a whole distributes the various faces it encounters to proprietary volumes, and subvolumes, and sub-subvolumes, according to the natural or objective similarities that variously unite and divide the many faces encountered during its training period.

It is thus important to keep in mind that the significance of the 80 distinct but elusively facelike patterns 'preferred' by the 80 middle or compression-rung cells is *not* that each pattern represents something in the objective world. These are *not* just so many 'grandmother cells.' Neither has the network simply 'memorized' all, or even a single one, of the faces in its training set. On the contrary, what is important about this learned set of preferred stimuli is that these 80 diagnostic templates provide the most effective armory for collectively *analyzing* any face, entered at the input layer, for subsequent placement in a well-sculpted *map* (viz., the second-rung activation space) of the important ways in which *all* human faces variously resemble and differ from one another.

From figure 2.13, you can see—even at a glance—that the network's second rung activation space, as shaped by the transforming synapses, has acquired a great deal of general information about *faces* in particular, as opposed to trees, or automobiles, or butterflies. Collectively (and *only* collectively), those 80 preferred stimuli represent both what the network now

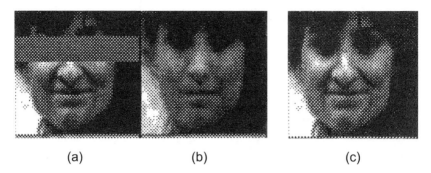

(a)   (b)   (c)

Figure 2.14
'Vector completion' (b) of a partial input image (a).

expects to find in its perceptual experience, and the network's resources for precisely *placing* any current input face within that framework of background expectations. The expression "expects to find" is substantially metaphorical, to be sure, at least at this simple level. But it also expresses a very important and entirely literal consequence of the compression-coding tactics here at work. Specifically, a partial or occluded version of a familiar input, such as the "blindfolded" face of Mary in figure 2.14a, will find itself reexpressed, at the output rung of a successfully trained compression network, in a form that simply *repairs* or *fills in* the missing portions of the input image, as in figure 2.14b.

For comparison, a photo of Mary from the original training set is included in figure 2.14c. You will notice that the eyes automatically interpolated by the network in figure 2.14b are not just any old pair of eyes: they are a fair approximation to *Mary's* eyes in particular. In speaking of what the network 'expects' to see, therefore, we are speaking about images that the network will *actually produce* at the output layer, even when the input data fall objectively and substantially short of specifying, on their own, the output in question.

The input deficit is made good, of course, by the network itself. Or rather, it is made good by the *general knowledge* about faces—Mary's face, and others—slowly acquired by the network during its training period. This capacity for leaping from partial data to a complete representational interpretation, called *vector completion*, is an automatic processing feature of any well-trained compression network. Variously degraded inputs will nonetheless be preferentially jammed by the resource-strapped network into the nearest of the available categories (i.e., the subvolumes of its middle-rung activation space) acquired by the network during training. For example, a further picture of Mary, unoccluded but partially corrupted by scattered noise this time, will likewise yield an output image of the same character and quality of figure 2.14b.

Plato would approve of this network also, because it, too, displays the capacity to reach beyond the shifting vagaries of one's sensory inputs so as to get a grip on the objective and enduring features of one's perceptual environment. Insofar, it also represents a capacity for what might be called "*structural* or *objectual* constancy" in the face of changing occlusions and perceptual noise, a capacity analogous, perhaps, to that for *color* constancy (in the face of variations in ambient light levels) displayed above by the Hurvich–Jameson network.

Though it does not make a leap from one *proposition* to another (but rather from one *activation pattern* or *vector* to another), vector completion is clearly a form of *ampliative inference*, no less than abduction, or hypothetico-deduction, or so-called 'inference-to-the-best-explanation.' Specifically, the accuracy of its representational outputs is not strictly guaranteed by the contents of its representational inputs. But those outputs are deeply and relevantly informed by the past experience of the network, and by the residue of that experience now resident in the structure of its well-trained middle-rung activation space. Accordingly, it is at least tempting to see, in this charming capacity for *relevant* vector completion, the first and most basic instances of what philosophers have called "inference-to-the-best-explanation," and have tried, with only limited success, to explicate in linguistic or propositional terms.

There are several reasons for taking this suggestion seriously. For one thing, there is no need, on this view, to construct any additional cognitive machinery to enable perceptual networks to rise to the capacity of 'ampliative inference.' The relevant vector-processing networks already have that capacity, and display it spontaneously, after training, in any nonchaotic domain. The cognitive capacity at issue is built into the basic network architecture, *ab initio*.

Second, the vectorial character (as opposed to a propositional character) of this capacity places it right at home in the brains of nonlinguistic animals and prelinguistic humans, who are presumably entirely innocent of any internal manipulations of propositional attitudes. Being thus at home, there is therefore no problem in ascribing such (vectorial) inferences-to-the-best-explanation to nonlinguistic creatures. They, too, can have sophisticated abductive cognition—as plainly they do—an achievement that is highly problematic on the classical and pointedly linguaformal approach to explanatory understanding.

Finally, the cognitive capacity here under discussion embodies an elegant solution to a chronic problem that bedevils the classical or lingua-formal approach to abductive inference generally. The problem, in brief, is as follows. What will count, for a given individual in a given situation, as the *best* of the explanatory takes or interpretations available, is a global function of *all* (or potentially all) of one's current background information. But short of conducting an exhaustive search of all of one's current background beliefs and convictions—an effectively impossible task, at least in real time—how can such an evaluation ever be relevantly or responsibly performed? Readers will recognize here what was called the "frame problem" within the classical or program-writing tradition in AI.

But it is better that you be reminded of this problem by a declared champion of the classical approach, rather than by one of its declared critics, so allow me to quote from one of Jerry Fodor's recent books (Fodor 2000), in which the problem here at issue takes center stage as the potential Grim Reaper for the hopes of the classical Computational Theory of Mind.

As for me, I'm inclined to think that Chicken Little got it right. Abduction really is a terrible problem for cognitive science, one that is unlikely to be solved by any kind of theory we have heard of so far. (42)

To say that a kind of inference is *global* [italics added] is to say inter alia that there's no bound on *how much epistemic context* [italics in the original] the rationality of drawing it may be sensitive to. (43)

Classical architectures know of no reliable way to recognize such properties [i.e., sensitivity to global information], short of exhaustive searches of the background of epistemic commitments. (38)

But let us now examine this (genuinely) vexing problem from the perspective of how a simple compression network comes up with an interpretive or ampliative representation, at both its middle and its output rungs, of a given sensory input. Fix again on the example of the blindfolded-Mary image as the input pattern, its 80-dimensional assay as the middle-rung pattern, and the completed-Mary image (fig. 2.14b) as the output pattern. Where, if anywhere, does the network's acquired *background* information reside? Answer: in the acquired weight-configuration of the 327,680 synapses meeting the second rung, and in the further weight-configuration of the 327,680 synapses meeting the output layer—roughly two-thirds of about a million information-bearing connections, even in a tiny model network.

Ask now, how much of that background information is causally effective and/or computationally deployed in transforming the occluded input image into its middle-rung format, and then into the relevantly completed output image? Answer: all of it. Every last synaptic bit of it. And, if the network were constructed of real neurons, this globally sensitive two-step transformation would take place in real time, because the first 327,680 computational steps (each one, recall, multiplies the strength of the incoming axonal signal by the weight of the connection at issue) all take place simultaneously, as do the second 327,680 steps, perhaps ten milliseconds later. While it is as 'globally sensitive' as one could wish, this robustly ampliative or abductive inference would be relevantly and responsibly completed in less than twenty milliseconds, that is, in less than 1/50th of a second.

Classical problem solved. And what solves it is the decidedly nonclassical architecture of a neural network engaged in massively parallel distributed *coding* (that is, representation by means of high-dimensional activation vectors) and massively parallel distributed *processing* (that is, multiplying such vectors by a proprietary matrix of synaptic weights to produce a new vector). Moreover, it simultaneously solves a second major dimension of the classical problem, namely, the evident *ubiquity* of abductive inference throughout the most humdrum aspects of our cognitive lives. Fodor's worries about abduction are not idle, and they are not new. They reach back to a concluding paragraph in a much earlier book (Fodor 1983), in which he expresses a well-conceived worry that the encapsulated computational activities found in his panoply of postulated 'modules' will prove unable to account for the vitally *un*encapsulated activities of his postulated 'Central Processor.' His worry was then supposed to be safely confined to that Mysterious Inner Keep. But his most recent book reflects Fodor's growing suspicion that abduction—and its hallmark sensitivity to enormous amounts of background or collateral information—is characteristic of cognition in general. Characteristic, perhaps, even of *perceptual* cognition! (Readers of my 1979 book will appreciate just how much I applaud Fodor's growing suspicions on this matter.) In the face of this increasingly evident ubiquity, a welcome feature of the vector-processing view of cognition outlined above is that globally sensitive abduction turns up as a characteristic feature of brain activity from its earliest and simplest stages, including its perceptual stages. And it remains central, as we shall see shortly, for even its most sophisticated activities.

It is important not to overstate this point. In a purely feed-forward network (i.e., one with no recurrent or descending axonal projections), the 'interpretive take' produced at any given neuronal layer or rung is sensitive only to the general information embodied in those cadres of synaptic connections that lie *at* and *below* that rung in the processing hierarchy. General information embodied at higher synaptic levels can make no contribution to any of the abductive steps that lie below them. But the point remains that, even for such a purely feed-forward network, each of these lower transformative steps is already a blatantly abductive step, an ampliative step that draws automatically and simultaneously from a vast register of antecedently acquired general knowledge about the domain addressed, in order to reach its interpretive 'conclusion.' That vast register is nothing other than the entire population of synaptic connections—pushing a million strong even in our simple three-rung models—that stepwise produced the interpretive activation pattern at the output rung.

Of course, as the original sensory input vector pursues its transformative journey up the processing ladder, successively more background information gets tapped from the synaptic matrices driving each successive rung. But the 'information tapping' mechanism is the same broadly focused, wisdom-deploying mechanism at every stage. Even a *purely feed-forward* network, therefore, is up to its neck in knowledge-sensitive or 'theory-laden' abductions, from its second rung representations on upward.

And when we lift our attention to massively *recurrent* networks, which biological brains most surely are, we behold a further mechanism for bringing background information to bear on each abductive step, a mechanism that makes each step at least potentially sensitive to *all* of the information embodied in the entire processing hierarchy, and not just the information embodied below it in that hierarchy. An exploration of this additional mechanism is the primary focus of chapter 4.

Fodor briefly turns to address, with more than a little skepticism, the prospects for a specifically 'connectionist' solution to his problem, but his discussion is hobbled by an outdated and stick-figured conception of how neural networks function, in both their representational and in their computational activities. His own reluctant summary (Fodor 2000, 46–50) wrongly makes *localist* coding (where each individual cell possesses a proprietary semantic significance) prototypical of the approach, instead of *population* or *vector* coding (where semantic significance resides only in the collective activation patterns across large groups of cells). And it wrongly assimilates their computational activities to the working out of 'associations' of various strengths between the localist-coded cells that they contain, instead of the very different business of transforming large vectors into other large vectors. (To be fair to Fodor, there have been artificial networks of exactly the kind he describes: Rumelhart's now-ancient 'past-tense network' [Rumelhart and McClelland 1986] may have been his introductory and still-dominant conceptual prototype. But that network was functionally inspired to solve a narrowly linguistic problem, rather than biologically inspired to address cognition in general. It in no way represents the mainstream approaches of current, neuroanatomically inspired connectionist research.) Given Fodor's peculiar target, his critique is actually correct. But his target on this occasion is, as it happens, a straw man. And in the meantime, vector-coding, vector-transforming feed-forward networks—both biological and artificial—chronically perform globally sensitive abductions as naturally and as effortlessly as a baby breathes in and out.

Moreover, allow me to repeat, such abductive transformations as do get performed, by the cadre of synapses at a given rung of any processing ladder, yield the materials for a further abductive inference at the next rung up. That in turn yields fodder for the next rung, and so forth, for however many rungs the network may contain, until the uppermost rung yields its own 'interpretive take' on the *already much-abduced* presumptive information that lands at its doorstep. Accordingly, the vaguely Platonic business of looking past the noisy and ephemeral appearances, to get a grip on the enduring reality behind them, is plainly an iterated process. It involves not one, but a succession of distinct abductive steps, only tens of milliseconds apart, each one of which exploits the relevant level of back-ground knowledge embodied in the peculiar cadre of synapses there at work, and each one of which yields a representation that is one step less stimulus-specific, one step more allocentric, and one step more theoreti-cally informed than the representation that preceded it in the processing hierarchy.

This functional arrangement suggests a partial explanation of the con-siderable intelligence differences across the animal kingdom. Perhaps the higher primates, and mammals generally, simply possess abductive pro-cessing ladders with rather more rungs in them. This could yield, for them, a more penetrating insight into the enduring categorical and causal struc-ture of the world. Alternatively, or additionally, perhaps the time-course of learning in those upper rungs is more delayed in its onset, and continues rather longer, after being engaged, in humans than in other creatures. That would afford us, in the end, both better-quality inputs on which to expend our eventual high-level learning efforts, and a longer period in which to exploit those high-level opportunities. As it happens, neuronal and synap-tic development in humans is known to be substantially delayed, relative to all other creatures, and it is maximally delayed for those of our cortical regions that are progressively further away (as measured in terms of the number of synaptic steps) from the sensory periphery.

Let me return to Cottrell's network. To this point I have discussed only the initial or auto-associative compression-net incarnation of that network, primarily because networks of that genre provide such a clear illustration of so many important lessons. But let me now discuss its second and final configuration (fig. 2.12b), wherein all of the hard-won synaptic weights meeting the middle rung are retained, frozen exactly as-is, but the top half of the network is simply thrown away, to be replaced by a small population of only eight 'detection cells,' met by a cadre of only (80 × 8 =) 640 *un*trained synapses.

The construction of this final configuration was motivated by the following questions. If the 80 cells at the compression rung (of the initial auto-associative network) now possess such a detailed grasp of the general structure of human faces, and of the eleven individual faces (e.g., Mary's) on which they were trained, can a new and quite different third rung be trained to discriminate, on the basis of the representation at the compression rung, whether an arbitrary input at the first rung is the image of a face or a *non*face? (That determination is to be the job of the first of the third-rung detection cells.) Can it be trained to distinguish between images of *males* and images of *females*? (That is to be the job of the second and third detection cells.) And can it be trained to reidentify, by means of outputting a five-digit binary 'proper name' ascribed to each of the eleven individuals in the training set, the *same individual* across distinct and variant photo-images of him/her? (This is to be the job of the last five cells in the new output rung.)

The answers are all affirmative. The first and last are not surprising. The network has a uniformly minimal response, across the entire second rung population, to any input image that is not a face (nonface images, of course, tend not to find close matches with any of the 80 vaguely facelike templates). This fact makes the face–nonface distinction an easy one for the output rung to learn, and its mature performance on such discriminations is virtually perfect. As for learning distinct names for each of the eleven individuals on which it was trained, this is also easy for a network that already codes their unique faces sufficiently well to reproduce them in detail, via vector completion, while in its earlier, auto-associative incarnation. It would have been surprising if the network had failed to learn this 'naming' skill, and its mature performance is again almost perfect.

By contrast, learning to make the male–female discrimination is a rather more intriguing achievement, since it means that the original coding strategy discovered by the middle rung, before the net was reconfigured in its upper half, had *already* become sensitive to the various but subtle differences that divide male from female faces. In other words, its sculpted activation space was already coding male and female faces in distinct and mutually exclusive subvolumes, subvolumes whose central or prototypical hotspots (see again fig. 2.3b and fig. 1.2) were plainly a fair distance apart, given the accuracy with which the new network now draws that distinction—98 percent correct within the individuals on which it was trained, and 86 percent correct across novel and arbitrary individuals. (And it achieved this subtle discriminatory capacity without any explicit instructions on which were the male and which were the female faces in the original training set.) Such a robust level of performance, though much less than perfect, is impressive for such a small network. Humans, when denied information about obvious secondary sexual characteristics such as hairdos, beards, and makeup, are only 92 percent accurate when assessing the gender of unknown people in uniform-format photo images.

When shown the faces of novel individuals (i.e., people outside the training set), the network was not, of course, in a position to assign a learned proper name to them, as it was for the faces in the original training set. It had never even seen these people before, let alone learned their names. But the network resolutely assigned an unambiguous proper name, anyway, to almost every novel face. And the name it assigned was typically the name of the person in the original training set that the novel face *most closely resembled*—that is, the person whose prototypical coding-point in the middle-rung activation space was geometrically *closest* to the coding point of the novel individual. Once again, we can observe the spontaneous tendency of any well-traveled network to assimilate novel stimuli to some one or other of the prototypical categories to which it has already been trained. It automatically falls into the 'best of its available interpretations,' of any input, relative to its acquired background knowledge. It engages, once more, in a globally sensitive abduction.

We can also see, in this example, how abductive inferences can often lead one astray, for the network's identification of these new individuals is plainly mistaken. They are, in fact, perfect strangers. People, and animals, make mistakes like this all the time. In fact, and motor slips aside, abductive or ampliative misinterpretations may be the most common sort of cognitive error that any creature ever makes. Given the feed-forward cognitive architecture that all terrestrial brains display, and given the ubiquity of abductive inference in the typical operations of that architecture, this is no surprise.

6 Neurosemantics: How the Brain Represents the World

We noted above that the face network was quite successful at recognizing the same individual across diverse photos thereof, photos with small differences such as facial expression, hair placement, glasses on or off, and so forth. The final output layer of the network's second-generation or 'diagnostic' incarnation (fig. 2.12b) contains five neurons that learn to display the appropriate five-element digital 'proper name,' a name assigned to the relevant individual during the training period. But describing that five-element output vector as a proper name is clearly a conceit, one that reflects *our* appreciation that there is a unique individual who is being encountered on different occasions. The network itself draws no such distinction, between a reappearing individual on the one hand and sundry instantiations of an abstract universal on the other. A distinction of that sophistication lies far beyond its feeble ken. As far as the network is concerned, *mary*, *gary*, and *janet* are just more general categories, no less than *male* and *female*, or *face* and *nonface*.

What we are looking at, then, in the mature network, is an acquired conceptual framework that displays a three-level hierarchy. The activation space of the second rung of neurons is divided, first, into two mutually exclusive and jointly exhaustive subspaces—*face* and *nonface*. The former subspace is similarly divided into two mutually exclusive sub-subspaces—*male* and *female*. The *male* region is further subdivided into six distinct regions, one for each of the six male individuals on which it was trained. And the *female* region is similarly subdivided into five distinct regions, one for each of the five female individuals in the original training set.

For dimensional reasons, I am unable to portray, graphically, the same sort of deformed grid-work that emerged in the two-dimensional space of our opening example (see again fig. 2.3b). But rest assured that the eighty-dimensional space of the face network's second rung now embodies a similar set of warped lines (Cottrell's network used neurons with a non-linear transfer function, like that portrayed in fig. 2.4). These lines are focused first and most gently on separating the coding region for nonfaces from that for faces-in-general. They are focused second and more closely on separating the two regions for male and for female faces, respectively. And they are focused finally and most tightly on the eleven subordinate regions for coding each of the eleven 'person types' (i.e., the eleven individuals) in the original training set. The mutual proximity of any two points within that space is a measure of the mutual similarity, as assessed by the network, of the two perceptual inputs that produce those two second-rung points. And distality therein is a measure of dissimilarity in the perceptual inputs.

That acquired categorical structure is a presumptive *map* of the domain of possible human faces—or at least, of that domain as-it-was-sampled by the small portion of it (eleven people) encountered during the network's training. That 80-dimensional map embodies a presumptive accounting of the most important ways in which faces can resemble and differ from each other. It also embodies a secondary accounting of the major groupings and subgroupings into which real-world faces tend empirically to fall. And it embodies a tertiary accounting of *where* those groupings are all located, relative to each other, within that carefully sculpted similarity space.

The metaphor of a map is not frivolous, and it is not superficial. A standard fold-out map of the interstate highway system of the continental United States provides a systematic accounting of the most important ways in which geographical locations can differ from one another: specifically, by being variously placed on a grid of equidistant *north–south* lines and *east–west* lines. It also embodies a secondary accounting of the general traffic-*cum*-population densities that distinguish the East and West Coasts from the more lightly traveled-*cum*-populated Central Plains. Specifically, the roadmap lines representing major highways are fewer and more widely separated in the center of the country, and they are more numerous and more closely pressed together as one approaches either coast, and most especially, as one approaches our major cities. And finally, since this two-dimensional map is a Euclidean manifold, the distance relations between the various city icons within that space, as measured in inches, are all proportional to the objective distance relations, as measured in miles, that hold between the objective cities themselves. (Given the curvature of the Earth, this isn't *exactly* correct, but as we shall see, few maps are.) Each city is portrayed as having a distinct and definitive *set of distance relations* to every other city on the map, and indeed, to every lake, river mouth, mountain, highway junction, or any other item portrayed on that map. Specify such a set of distance relations, and you have uniquely specified the city that boasts them.

Similarly, if we specify, metrically, the set of similarity and dissimilarity relations that an arbitrarily chosen face-type (i.e., individual) bears to the other ten face-types (i.e., individuals) in the network's training set, then we have uniquely specified the face-type that boasts them: Mary's, for example. This fact provides the opportunity for the student network to get a genuine grip on Mary's face—indeed, on all of the training faces—by slowly reforming the response profile of its second-rung activation space so that the proximity and distance relations, within that space, of the emerging eleven face-coding regions, come to be collectively proportional to the objective facial similarity and dissimilarity relations displayed within the training population.

In this case, of course, the objective domain being portrayed is not a space of possible geographical locations, as with the highway map. It is a space of possible *human face configurations*. And in this case, of course, the portrayal itself is not a two-dimensional sheet of printed paper. It is an *80-dimensional neuronal activation space*, carefully and relevantly sculpted by many epochs of training on the objective domain at issue. But applying the expression "map" has exactly the same rationale in this case as it does in the case of the road map. Specifically, there is a set of salient points (the centers of the eleven densely used coding regions for the eleven face-types) within the map itself, and a set of salient items or categories in the world (the eleven individual faces, or strictly, 'face types'), such that the assembled (high-dimensional) distance-relations between the salient points within the map accurately mirror (are homomorphic with) the assembled similarity relations that hold between the target items/categories in the world.[7]

Thus does the brain represent the world: not one category at a time, but an entire *domain or family* of intricately *related* categories at a time. And it does so by means of domain-portraying maps that contain, by virtue of their acquired internal structure, an enormous amount of systematic information *about* the domains portrayed, and about the categories within their embrace. That acquired information displays itself, or some relevant part of itself, every time it funds an abductive inference, which is to say, every time the network does anything at all.

These objective domains can be as diverse as you please, because the sundry vector-space maps that come to represent them need only address the abstract metrical structure of the peculiar similarity-and-difference relations that characterize the peculiar family of features or properties within the domain being represented. Thus, a vector-space map (i.e., a well-sculpted neuronal activation space) can successfully represent the domain of human faces, as we saw. Or, such a map might be differently trained, so as to represent the domain of *human voices*: perhaps with an

7. This approach to the nature of representation within the brain is also defended in two recent papers by G. O'Brien and J. Opie, a psychologist and philosopher, respectively. See O'Brien and Opie 2004, 2010. But the basic idea goes back a long way; see Shepherd 1980.

initial division into male, female, and children's voices; and then with appropriate subdivisions into tenors, baritones, and basses for the males, and into sopranos, mezzos, and contraltos for the females. A third map, exploiting the same strategy of 'relevant homomorphism,' might represent the voices of the various *musical instruments*: dividing that domain, first, into the strings, the brasses, and the woodwinds; and subsequently into violins, violas, cellos, and basses for the strings; and so forth for the other two categories. Alternatively, a single map might encompass both the human voices and the instrumental voices, with each group confined to its own subspace. A fourth and rather larger vector-space map—within the brain of a professional biologist, perhaps—might address the entire domain of *animals*, and achieve a level of hierarchical complexity equal to the entire classification system that we owe to Linnaeus. A fifth map might portray the domain of *heavenly bodies*, either with a geocentric conceptual system like that of Aristotle and Ptolemy, or with a heliocentric system like that of Copernicus, or with a system that ascribes no center at all to the universe, like that of modern astrophysics.

With these last two examples we may begin to observe that different networks, within different individuals, perhaps within different cultures or stages of history, can embody importantly *different* maps of one and the same domain of phenomena. They can, and regularly do, embody quite different *conceptions* or *understandings* of the objective domain jointly addressed. They can differ in the level of detail and complexity that they achieve, as an adult's vector-space map of almost any domain will exceed a child's map of the same domain. And they can differ in the levels of accuracy and penetration that they achieve, as when the superior map correctly embodies the real dimensions of similarity and difference that unite and divide the various categories embraced, while the inferior map wrongly structures itself according to similarities and differences that are merely superficial, incomplete, perspective-dependent, or confused. Think again of the examples of Ptolemaic, Copernican, and modern astronomies. We shall return to this topic later in the book.

We are here contemplating a novel account of the meaning or semantic content displayed by human and animal concepts. I have previously referred to it as "State-Space Semantics," for obvious reasons, but it might more aptly be called "Domain-Portrayal Semantics," for that is what our various maps do: portray an abstract feature-domain. It is important to contrast it with the major alternative accounts held out by the philosophical tradition. We scouted some of the differences, briefly, in the opening chapter. Let us now look into the matter more deeply.

7 How the Brain Does *Not* Represent: First-Order Resemblance

On the account of concept meaning held out by Locke and Hume, our basic or simple ideas are faint copies of the various simple sensations produced as we encounter the various sensible features of external objects. Straightforward *resemblance* is the fundamental semantic relation: your elementary idea or simple *concept* of straightness resembles your *sensation* of straightness, which in turn resembles the external *property* of straightness displayed by rods, ruled lines on paper, stretched strings, and light rays. Complex concepts are created by modulating and concatenating the simple concepts, but even with complex concepts, resemblance remains the fundamental semantic relation. For the manner in which the simple concepts are modulated and concatenated, in a given complex concept, is a direct reflection of the way in which the simple objective properties are modulated and concatenated to comprise the complex external object. Complex concepts that fail to find any such objective resemblances are concepts of chimerical, or nonreal, objects, such as unicorns and griffins. The constituting elements of those concepts may denote real properties, but their peculiar assemblages in these two concepts correspond to no similar assemblages of properties in the real world.

Moreover, even certain *simple* concepts may be judged devoid of external reference, if they fail the test of resemblance between the external property represented and the intrinsic character of the concept that supposedly represents it. Recall Locke's decision to deny colors, tastes, and sounds the status of objective or 'primary' properties, on grounds that modern science has revealed that external bodies possess no objective properties that genuinely *resemble* the qualitative characters of the peculiar sensations and ideas that those external bodies regularly produce in us. Failing any such legitimating resemblances, the various colors and sounds, and so on, must therefore be demoted to the status of merely 'secondary' properties, properties that have no instantiation save, perhaps, as features of our internal cognitive states.

This familiar position may seem just a welcome recognition of the general point that reality may not be, in all respects, exactly as we perceive it to be. But in fact, the theory of semantic content on which it is based is fatally flawed in any number of respects. For one thing, the initially modest Lockean irrealism, the one motivated by the resemblance account of semantic content, threatens to slide immediately down a slippery slope toward an entirely immodest blanket Idealism. Berkeley, eyeing Locke's deliberately limited irrealism, decided that our ideas of so-called 'primary' properties were no better off, resemblance-wise, than are our ideas of the so-called 'secondary' properties. And on the strength of this, he quite understandably decided to write off the corresponding ontology of objective, mind-independent physical objects, with their alleged properties of objective shape, objective motion, and objective mass, just as Locke had written off the colors, tastes, and sounds. Accordingly, on Berkeley's view, the physical world is unreal in its entirety. The only substances left were nonphysical minds—God's, and our own—and the only properties left were properties of minds and their internal states.

I am not approving of Berkeley's fabulous conclusion here. But he did have a dialectical point. Locke is clearly going to have a difficult time drawing a clear and well-motivated distinction between those cases where our concepts do genuinely resemble the external properties they supposedly denote, and those cases where they do not. A salient case in point is one of Locke's presumed *primary* properties: an object's *mass*. It is not at all clear to me that my concept of mass literally *resembles* the objective property of mass in any respect at all. Indeed, I'm not sure what sort of resemblance I should even be looking for here. On Locke's criterion, then, perhaps mass should be tossed out of our considered ontology along with the colors.

Even our concepts of shape and motion are suspect on Locke's criterion of successful objective reference. Most famously, Immanuel Kant erected a second Idealist edifice, a subtle alternative to Berkeley's, in which even spatial and temporal properties were denied instantiation in the 'Things as they are in Themselves' that somehow lie behind the familiar 'Empirical' world—the world-as-we-think-and-perceive-it. Kant also fought the assimilation of his own view to Berkeley's unabashed Idealism, first, by insisting on the literal integrity (suitably understood) of our familiar idioms for distinct material substances within an objective empirical world, and second (and more importantly), by insisting that the ontology of our inner or *mental* world is no less a domain of mere appearance as well. But Kant remains an unrepentant Idealist, if only from the austere (and highly problematic) 'transcendental' point of view. And from that latter point of view, the 'resemblance gap' between our concepts and 'Things in Themselves' is here made even wider than Berkeley supposed. As Kant saw it, we cannot even speak, sensibly, about resemblances across that abyssal divide.

The way to avoid the irrealist or Idealist bottom of Locke's slippery slope (see fig. 2.15, plate 8) is never to step on that slope in the first place. A confused rationale for so refusing would be to adopt a blanket Naïve

Figure 2.15
The slippery slope to Idealism (my thanks to Marion Churchland). See plate 8.

Realism from the outset: try to insist that *all* of our perceptual ideas find legitimating resemblances to external properties. Aside from being indefensible, this response misses the point. The issue is not whether or which of our ideas succeed or fail in finding such resemblances. The issue is whether resemblance is the relevant semantic relation in the first place.

A better reason for refusing to take Locke's initial irrealist step is simply to reject his resemblance-based standard for the reality of objective perceptual properties. By this, I do not mean that we should never question the integrity or objective purchase of our current concepts. Of course we should—even our perceptual concepts. We have done so successfully in the past, and we will do so again in the future. Think of our ancient concepts of 'up' and 'down,' for example, or our perception of the Sun's 'daily circular motion.'

But the criterion for *evaluating* the integrity or objective purchase of any concept should not be Locke's naïve and ill-considered criterion of a one-to-one concept-to-property resemblance. That would be like requiring, of a highway map of the entire continental United States, that the various circular black dots for the cities of Chicago, Philadelphia, San Diego, Houston, Seattle, and New York must each literally *resemble* the unique city that each black dot represents.

On a continental highway map there is, of course, *no* such direct or first-order resemblance between the representing circular black dot and the city it represents, nor *need* there be for successful representation. Instead, successful representation is achieved—again, of course—by the very different *second-order* resemblance between the overall pattern of distance-relations between the various circular black dots on the map, and the overall pattern of objective distance relations between the various cities collectively represented. More generally, anything on a continental highway map—a red line for a highway, a blue line for a river, a black line for a state-to-state border, a curving pale-green to pale-blue border for a coastline—has the objective reference or denotation that it does because of (1) the family of *distance-relations* it bears to everything else on the map, and (2) the homomorphism that holds between those assembled distance relations and the objective distance relations holding between the objective items that are thereby successfully mapped. Colors can be part of a useful iconic code, to be sure. But the highways themselves are not red; the rivers are not remotely blue; state lines are not black—indeed, they are entirely unmarked, save for the occasional welcome sign; and a real coastline is not a pale-blue against a pale-green boundary. The resemblance that constitutes the map's representational success is a collective second-order or abstract structural resemblance, rather than a large number of independent first-order resemblances.

Evidently, the proper semantic theory for a highway map, or any other geographical map, is robustly *holistic*. Semantic significance accrues to such map elements not one by one, and independently of the semantic significance of neighboring map elements. Rather, they gain their semantic and referential significance collectively, and the unique significance of each distinct element is an irreducible function of all of the *within-map* relations it bears to every other map element. For maps, at least, no other semantic theory is even worth talking about.

Now, I admit that establishing the claim that any creature's conceptual framework is most usefully seen as a high-dimensional vector-space map is one of the burdens of this book, and it is a burden not yet fully discharged. But you may observe this claim hard at work in the preceding discussion, as providing an explanation for, and a motivated escape from, the variously agonized, extravagant, and entirely unnecessary irrealist and Idealist posturings of Locke, Berkeley, and Kant. In sum, if we opt for *first-order resemblance* as the primary semantic relation between concepts and objective properties, we will find ourselves oscillating unstably between a blanket Naïve Realism on the one hand, and a blanket Idealism on the other. Neither extreme is acceptable. Nor is an unstable oscillation between them. So let us jettison the semantic account—first-order resemblance—that forces us into this situation.

There are at least two further reasons for rejecting the Locke–Hume account of what concepts are, and of how they gain their semantic significance. Perhaps the most obvious reason concerns the origins or *genesis* of our so-called 'simple ideas.' They are copies, we are told, made when, and only when, the mind encounters an instance of the corresponding 'simple impression' or sensation. This story thus assumes that our sensations also admit of a determinate decomposition into an antecedent alphabet of 'simples,' simples that correspond, finally, to an antecedent alphabet of 'simple' properties in the environment.

But this is a child's view of the situation, both as it concerns the constitution of the external environment, and as it concerns our internal mechanisms for its cognitive representation and the development of those mechanisms over time. It might have been residually plausible in the seventeenth century, but it is not remotely plausible here in the twenty-first.

Let us begin with the external environment. If there are any 'simples' underlying its manifest complexity, they are features such as energy, mass, electric charge, conductivity, specific heat, magnetic field strength, wavelength, condensation point, melting point, and the 92 nucleon-plus-electron-shell configurations that constitute the 92 natural elements with their characteristic chemical valences. But these basics from physics and chemistry are not the 'manifest' properties that answer to our workaday perceptual concepts. It has taken millennia of determined scientific research to unearth and appreciate the mere existence of such basic properties.

On the other hand, such objective properties as *are* detected or tracked by our native sensory organs are typically very complex properties indeed.

For example, one's ears track, across four orders of wavelength magnitude, the intricate power spectrum of incoming atmospheric compression waves, and the various directions from which its diverse components arrive to one's head. One's tongue tracks the comings and goings of thousands of different kinds of molecules in solution in our saliva, some of them fairly simple such as the ions of sundry acids and bases, but many of them rather complex such as the carbohydrate structures of the several different sugars, and some of them extremely complex such as the endlessly various protein molecules that make up so much of what we eat. It also tracks the temperature, the viscosity, and the mechanical textures of whatever materials enter one's mouth.

One's nose performs an analogous task for the endless varieties of molecules, large and small, in atmospheric suspension. The somatosensory system(s) within one's skin track a wide variety of mechanical, thermal, chemical, and radiative interactions with the local environment, not all of them easily distinguishable, even from one another. And finally, one's eyes confront perhaps the greatest complexity of all: a three-dimensional jumble of opaque physical surfaces and semitransparent media, each one occluding, absorbing, reflecting, refracting, diffracting, and scattering the ambient electromagnetic radiation according to its own idiosyncratic molecular makeup. Contrary to the Locke–Hume picture at issue, the empirically accessible objective world does not come precarved into obvious simples. Rather, it presents itself initially as (to borrow from William James) a "blooming, buzzing confusion" in desperate *need* of subsequent partitioning into predictively and explanatorily relevant categories.

That subsequent partitioning is precisely what the process of synaptic modification is bent toward achieving. For the various one-step transductions performed at the absolute peripheries of our several sensory modalities largely *repeat* the real-world complexities just noted, leaving the job of performing progressively more useful categorizations to the synaptic matrixes and the neuronal populations at the rungs successively higher in the information-processing ladder. *Color*, for example, is not explicitly represented at the level of one's retinal cone-cells. As we saw earlier, even in the simple Hurvich network, color is not explicitly represented until it finds expression within the *second*-rung population of the model's opponent-process cells. And in primate brains, it may well find its principal representation, even as a 'purely sensory' property, in cortical area V4, an area rich in color-opponent cells some six or seven rungs (i.e., synaptic steps) away from the cone population at the retina (Zeki 1980). This does not mean that wavelength information goes utterly unused before that stage. But it does mean that our familiar color categories go entirely unrepresented at the very first level of our sensory activities.

As well, and to return to Locke, the preservation of *resemblances*, as one climbs the processing ladder from sensory neurons through successive neuronal populations, is the last thing with which the brain is concerned. The point of successively 'reformatting' what begins as a sensory activation pattern is precisely the successive *transformation* of that confused, entangled, and perspectivally imprisoned information. It gets transformed into a series of representations that increasingly reflect the brain's acquired or background information about the enduring spatial, categorical, and causal structure of the objective world, the world that originally gave rise to that fleeting and highly idiocentric sensory input.

A final defect of Concatenative Empiricism is the false picture it provides of the temporal order in which concepts are standardly learned, and the internal structure that those concepts are supposed always to enjoy. The alleged copying process that leaves a simple idea behind after a simple impression has come and gone would suggest that it is some preferred set of simple ideas that gets learned first, and only subsequently are they variously concatenated, by the mind, to form the recursively constituted 'complex' ideas. But in fact, the 'entry-level' concepts initially learned by toddlers are what the contemporary tradition would call *mid*level concepts such as cookie, doggie, shoe, mailman, and spoon (see Anglin 1977, 220–228, 252–263). Only subsequently do they learn anything that might be a constituting subfeature of these midlevel categories, and then only slowly. The color categories, for example, are only rarely learned by most children before their third birthday, after they are already fluent in a rich vocabulary of complex practical, social, and natural kinds, and have been for fully a year or two. Evidently, the colors are a fairly late abstraction from an already complex conception of the world, not a 'given simple' from which that complex conception is subsequently constructed.

As well, neither our color sensations nor our color concepts are plausibly counted as 'simples' in any case. Each color sensation is a three-dimensional activation vector, however ignorant one may be of that internal structure. And our late-blooming color concepts, like all other concepts, are inescapably embedded in a framework of general knowledge about what kinds of things typically display which colors in particular, and about the typical causal properties that they will thereby possess. Recall once more (cf. the face network) the wealth of *general background knowledge* about its target domain that any network inevitably acquires in the process of getting a grip on the categorical structure of that domain.

The origin of simple ideas and the temporal order of learning aside, the great majority of our 'complex' concepts stoutly *resist* a recursive semantic reconstruction in terms of concepts that might plausibly be counted as semantic 'simples' in any case. Try defining *democracy*, or *war*, for example, in terms that meet this condition. For that matter, try doing it for *kitten*, or even *pencil*. Anyone who undertakes this task learns very quickly that it produces only confusion and disappointment.[8]

From the perspective of a state-space or domain-portrayal semantics, the explanation for this reconstructive failure is quite simple. The neuronal activation space in which any human concept is located is typically a very high-dimensional space, with as many as perhaps a hundred million dimensions. All of them are important, but each one of those dimensions contributes only an insensible and microscopic part of the all-up profile that registers the concept as a whole. Those dimensions are, and are likely to remain, utterly unknown to the cognitive creature who deploys them. One should not, therefore, *expect* that our concepts can generally be (re)constructed from a finite class of obvious simples. For on the sculpted-activation-space view, *all* of our concepts are 'complex,' though not at all in the ways that the Locke–Hume tradition assumed. Like the elements of any map, they all derive their unique representational content or semantic significance from the unique weave of distance and proximity relations they bear to every other element of the enveloping map. And that weave of relations is embodied and sustained in a space of a great many dimensions.

A mature conceptual framework often displays a hierarchical structure of some kind or other, to be sure, if not of the kind just rejected. The acquired framework of the face-recognition network provides a straightforward illustration of this fairly common outcome. The subordinate categories for the five individual face-types *mary, janet, liz, jean*, and *pat*, for example, are all subvolumes of the larger activation-space volume for *female faces*. That larger volume is a subvolume in turn of the still larger volume for *human faces*, a volume that is exhausted by the disjoint subvolumes for *female faces* and *male faces*. This arrangement reflects the network's implicit (repeat: *implicit*) knowledge that Mary, Janet, etc., are all females, that all female faces are human faces, and that no females are males.

But no semantic 'simples' jump out at us here, begging to be recognized as the basic elements in a recursively constructed hierarchy. One might

8. I was intrigued to find that even Fodor is on board for this claim, with important arguments of his own. See Fodor et al. 1985.

hope to find them in the constituting *axes* of the network's middle-rung activation space—that is, in some individual 'semantic contents' of the individual neurons at the middle-rung population—but that hope is quickly exploded. As we saw earlier in this chapter, if one reconstructs the 'preferred stimulus' for any one of the 80 cells at that rung, something that is easily done from the acquired weights of the 4,096 synapses arriving to it, one finds that the unique input stimulus that produces the maximal response at that cell is typically a diffuse, elusive, but variously facelike pattern that spans the entire 'retinal' surface (see again the six examples in fig. 2.13). These complex templates are not 'simples' in any sense of the term. They concern intricate spatial structures across the entire retinal surface; they have fully 4,096 constituting pixels; and they correspond to nothing simple in the environment.

Indeed, they correspond to *nothing at all* in the network's actual perceptual environment, for none of them is a 'picture' of anyone the network has ever encountered. Instead, those 80 different diffuse patterns slowly emerged in the course of training because *collectively* they constitute the most useful set of distinct similarity-templates (or the most useful set that the back-propagation algorithm could *find* during that training run) for analyzing the differences and similarities that divide and unite the domain of human faces generally. That is why the network performs so well in its discriminatory activities, even for novel faces (i.e., faces outside the original training set).

We may both underscore and illustrate the profoundly nonclassical, nonconcatenative role played by the many axes of the network's second-rung activation space, by pointing out that if we train 100 distinct networks to the same level of discriminatory skill, and on the same set of faces, we will likely find that all 100 of the networks have settled into a set of activation-space partitions very close in its internal structure to that portrayed in figure 1.4. But *no two of them* need share even a single 'preferred stimulus' of the sort portrayed in figure 2.13. Each network will have its own idiosyncratic set of such variously facelike diagnostic templates.

That is because the structured set of partitions and similarity relations, though shared across all 100 of the trained networks, *is differently oriented* within the 80-dimensional second-rung activation space of each and every one of them. It is variously rotated, and perhaps mirror-inverted, along any number of different internal axes, from network to network. For what matters to the diagnostic layer, at the third and final rung of each network, is not the absolute location of any second-rung activation point relative to the second-rung *neuronal axes*. It is the location of that second-rung activation point relative to all of the other *learned partitions and prototype points* within the second-rung space. For *that* is the acquired structure to which the third-rung diagnostic layer in each network has slowly become tuned, and which alone allows it to make the accurate output discriminations it does. That final layer knows nothing, and couldn't care less, about how that vital structure happens to be *oriented* relative to its sustaining axes at the second rung. That information is irrelevant to each and every network. So there is no surprise that it is not uniform across all of these networks (see Cottrell and Laakso 2000).

In sum, it is simply a mistake—a misconception of how neural networks do their job—to construe the neuronal axes of any activation space as representing anything remotely like the classical sensory 'simples' of the Locke–Hume tradition. And it compounds the mistake to construe the admittedly complex concepts that the network does learn as somehow *constituted* from those (nonexistent) simples. To be sure, the preferred stimulus that does answer to a given neuron plays a small but important *epistemological* role in helping the network to *identify* any perceived instance of the categories it has learned. But such microcontributions to an epistemological process will vary, and vary noisily, across distinct but similarly competent networks.

What does not vary across those networks, and what carries the burden of their shared semantic, conceptual, or representational achievement, is the shared structure of similarity and distance relations that unite and divide their acquired family of partitions and prototype regions. If you are interested in individual and idiosyncratic *microepistemology*—in what makes the appropriate point in some network's vector-space map *light up* on a given occasion—then you may well look to the 'preferred stimulus' of each cell in the relevant neuronal population. But if you are interested in *semantics*—in what bestows a *representational significance* on any given point—then look to the set of wholly internal relations that embrace, and uniquely locate, every prototype-element within that enveloping vector-space map.

I here emphasize this fundamental dissociation, between the traditional semantic account of classical empiricism and the account held out to us by a network-embodied Domain-Portrayal Semantics, not just because I wish to criticize, and reject, the former. The dissociation is worth emphasizing because the latter has been mistakenly, and quite wrongly, *assimilated* to the former by important authors in the recent literature (e.g., Fodor and Lepore 1992, 1999). A state-space or domain-portrayal semantics is there characterized as just a high-tech, vector-space version of Hume's old concept empiricism. This is a major failure of comprehension, and it does nothing to advance the invaluable debate over the virtues and vices of the competing contemporary approaches to semantic theory. To fix permanently in mind the contrast here underscored, we need only note that Hume's semantic theory is irredeemably *atomistic* (simple concepts get their meanings one by one), while domain-portrayal semantics is irreducibly *holistic* (there are no 'simple' concepts, and concepts get their meanings only as a corporate body). Any attempt to portray the latter as just one *version* of the former will result in nothing but confusion.

Still, and with the preceding dissociation acknowledged, we may ask if there is really no place at all, within a Domain-Portrayal Semantics, for the sorts of 'constituting' properties that both common sense and the philosophical tradition have been inclined to discover—properties such as *pink*, *slender*, and *beaked* relative to the more complex property *flamingo*. Indeed there is a place for such properties, and there is no harm in assigning an objective constitutive role to them. But the order of our concepts need not always follow or mirror the order of objective properties. A two-year-old child, well-schooled by frequent visits to the local zoo, may have a robust and distinct concept of flamingos and be highly accurate in their spontaneous identification. (My own son, at that age, called them "mingamoes!" and always took loud delight in thus announcing their presence.) And yet the same child will have no distinct and functioning concept for either *pink*, *slender*, or *beaked*. Those emerge only later, as the child slowly learns that such features are useful dimensions of analysis for a wide variety of his complex categories, well beyond the peculiar case of flamingos.

There should be no surprise at this 'lack of decomposition.' It reflects the empirical course of normal concept learning long familiar to developmental psychologists. And a similar lesson for such naïve Humean expectations emerges from the behavior of Cottrell's face-recognition network. While, arguably, it has a fairly sophisticated concept for *female face*, one that displays substantial background knowledge and breadth of application, the network has no distinct concept for *nose*, *eyebrow*, *mouth*, *chin*, or *pupil*. Nor has it acquired any skills in identifying such things when they are presented in isolation from the other face elements. Bluntly, the network has the complex concept of (that is, a metrically compacted prototype region for) *female face* without having an explicit concept for any of the constituting elements that actually make up objective female faces.

What it may well have, however, is a small handful of variously tilted hyperplanes within its 80-dimensional middle-rung activation space, hyperplanes within which certain factors *relevant* to being either a male face or a female face are preferentially registered. For example, among the many physical differences that are objectively relevant for distinguishing male and female faces are (1) the vertical distance between the pupil and the eyebrow (it is usually larger in females than in males), and (2) the distance between the bottom of the nose and the tip of the chin (it is usually larger in males than in females). (See again fig. 1.6 for an illustration.) Given the scattered variety in the preferred stimuli acquired by the 80 cells at that rung, some smallish number of them—say a dozen—may well be united in having *part* of their activation-level responses driven by the objective nose-to-chin distance of the current face image at the input layer, while otherwise their response-behaviors are uncorrelated. The activation patterns across the 12-dimensional subspace comprehended by those 12 cells are therefore a statistically relevant index of nose-to-chin distance, an index to which the output layer can become preferentially tuned in the course of training.

The same thing may happen, for a different group of cells, with regard to pupil-to-eyebrow distance; and so on, for other possible male–female discriminanda. In the end, the all-up 80-dimensional activation space at the middle rung of the matured network is shot through with various tilted subspaces, each of which is statistically tied to some relevant dimension of objective variation, and each of which contributes its own modest contribution to a collective discrimination that, in the end, is a highly reliable determination of gender.

But note well, nothing requires that these underlying dimensions of evaluation are *the same* from one network to another, or from one individual to another. They may well be. But they need not be. There are many different ways to get an epistemological grip on any given objective category, and different individuals may develop different dimensions of analysis. A color-blind person, or an alligator (also color-blind), may be as reliable as anyone else at identifying flamingos visually, but the color pink will play no role in the analyses of either creature's concept of that bird. Indeed, for some creatures, visual features may play no role at all. A blind bat, for example, may know what flamingos are, and be perfectly able to discriminate flamingos from other flying creatures, but it will do so by focusing (actively) on the distinctive acoustic signature of the sonar echo returned from such a bird, or (passively) on the equally distinctive acoustic signature ("swoosh, swoosh") given off when any flamingo takes flight. Even for networks or individuals who do share the same 'dimensions of analysis' for a given category, those dimensions may be very differently *ranked* in their relative causal importance to the perceptual discriminations that each individual makes. Hence, for a sample of faces sufficiently large to contain some ambiguous or borderline cases, no two networks, and no two people, will make *exactly* the same judgments as to the gender of every face in the sample.

Accordingly, we can account, in a natural and empirically faithful way, for the various intuitions behind the Locke–Hume account of concepts. But we should not be tempted by any aspect of the account itself. We can now do much better.

8   How the Brain Does *Not* Represent: Indicator Semantics

A popular contemporary alternative to the Locke–Hume account of concept meaning is the *indicator semantics* championed by Fodor, Dretske, Millikan, and others. It comes in different versions, some of them resolutely nativist and some of them more empiricist in inclination, but these differences are relatively unimportant for our purposes. What these diverse versions all share is the idea that the semantic content of any given mental representation *C* is ultimately a matter of the nomic or causal relations that *C* bears to some specific feature in the environment, relations that bestow upon *C*, or more strictly, upon tokenings or activations of *C*, the status of a *reliable indicator* of the local presence or instantiation of the objective feature in question. The concept *C* thereby acquires *that objective feature* as its proprietary semantic content.

Thus, a specific height—say, three inches—of the column of mercury in a bedside thermometer has the specific content, *98.6 degrees Fahrenheit*, because that particular height is a reliable indicator of exactly that temperature in whatever medium surrounds the bottom bulb of the thermometer. And the angular position of the needle on a voltmeter—pointing at "3 V," say, has the specific content, *three volts*, because that particular position is a reliable indicator of exactly that potential drop across the circuit elements being measured. And so forth for all of the other specific heights and positions on both of these measuring instruments. The full range of possible instrument behaviors corresponds to a range of possible objective features, and the instrument's current output behavior registers which specific feature, from the range of possible objective features, currently obtains. Our native sensory organs and perceptual systems are thus brought under the umbrella of measuring instruments generally (which is surely where they belong, their natural or evolutionary origins notwithstanding); and a positive contact with epistemological theory is thereby made, right at the outset.

There is something profoundly right about the view just outlined, but it fails utterly as an account of how or where a concept gains its *semantic content*. The underlying nature of that failure is quickly illustrated (though not strictly demonstrated) by looking once more at the source of the semantic contents of the elements of a continental highway map, or of an urban street map.

A standard, fold-out paper map also represents a range of possibilities—namely, a two-dimensional manifold of possible geographical positions that you, or any other physical particular, might occupy. But such a map differs from a typical measuring instrument because it lacks any *causal interaction* with the environment that would register, on the map itself, *where* in the continental United States, or in the urban street grid, the map and its reader are currently located. One can easily imagine such an indicator, to be sure: the brilliant dot of a laser pointer, for example, flanked by a luminous "You are here," that slowly glides across the surface of the map in perfect sympathy with the geographical movement of the map itself across the physical domain that it represents.

In fact, we no longer need to imagine such an unusual and almost magically informative map, because modern technology has already produced them. Upscale automobiles can now be purchased with an onboard GPS (Global Positioning System) coupled to a dashboard map display with an upward-pointing car icon (in plan view) at its center. As the automobile itself drives around the local roads, the relevant portion of the computer-stored map is automatically displayed underneath the fixed car icon, and its map elements flow by that icon in sympathy with the corresponding road elements flowing past the automobile. Watching the unfolding dashboard display, as you drive around, is rather like watching your automobile from a trailing helicopter several hundred feet above and behind you.

Thanks to the automobile's electronic linkage to the several geostationary GPS satellites in orbit above it, its computer-stored and computer-displayed road map has now been upgraded to a continuously operating *measuring instrument*, one that reliably registers or indicates the current geographical location of the automobile (and the on-board map) itself. The elements of the background map now meet the conditions, for semantic content, required by the Indicator Semantics approach outlined above. Accordingly, we can now safely and justly ascribe, as the proprietary referent or semantic content of each map element, the objective geographical elements or locations of which they (or more strictly, their current illuminations) are now such reliable indicators.

But hold on a minute. Those map elements *already* had those objective geographical elements or locations as their proprietary referents or semantic contents, prior to and wholly independent of any GPS linkage or shifting "You are here" illuminated display, and they would still possess them even if that causal link were disconnected. Indeed, the map elements of *any* standard, fold-out highway or street map have all of the relevant references, semantic contents, or representational status without the benefit of any position-indicating causal connections whatsoever. In short, a traditional road map is not a dynamically active measuring instrument at all, and yet it is a paradigm of a systematic and successful representation.

So there must exist at least *one* alternative route by which a representation can acquire an objective reference or semantic content—an alternative, that is, to the causally connected indicator route proposed above. The alternative is not hard to find, nor is it difficult to understand. It is the holistic, relation-preserving mapping technique displayed, for example, in any roughly accurate highway map.

Still, it requires an argument, above and beyond the observations of the last few paragraphs, to establish that *human* concepts in particular acquire their semantic contents in this alternative way. It is an abstract possibility, perhaps, but hardly a necessity. What can we say then, about the relative virtues of Domain-Portrayal Semantics over Indicator Semantics?

Quite a bit. Perhaps the first fact to demand our attention is that only a smallish subset of our concepts-at-large have the status of *observational* or *perceptual* concepts, in anything like the manner required by the familiar measuring-instrument analogy. The great bulk of our concepts find their singular applications, not in a spontaneous and cognitively unmediated response to one's local environment, but as the result of some fairly sophisticated and entirely defeasible inferences—either deductive or abductive—from the general fund of information available to one. Accordingly, the Indicator Semantics approach is obliged to assign a privileged status to the semantic contents of our perceptual or observational concepts in particular, and only a secondary or derivative status to the semantic contents of our many nonperceptual concepts. And it is further obliged to provide an account of how the semantic contents of the latter are somehow to be conjured out of the semantic contents of the former.

We have been here before. The Locke–Hume approach to semantic content also embraced a privileged class of explicitly perceptual concepts, and confronted almost exactly the same problem. As we saw earlier, the poverty of its proposed solution (successive concatenative definitions) is one of its most embarrassing failures. And the advocates of the Indicator

Semantics approach here at issue bring no salient new resources with which to address this problem. The question of the semantic contents of our many *non*perceptual concepts remains, for them, an unsolved and evidently long-festering problem.

For Domain-Portrayal Semantics, by contrast, there is no problem to be solved; the semantic gulf that wanted bridging above is here never opened in the first place. On our view, nonperceptual and perceptual concepts are on exactly the same footing, so far as the source or basis of their semantic contents is concerned. And that source is the relative position of the concept in a high-dimensional similarity space that also contains a large number of *other* concepts, which *family* of concepts collectively portrays (or purports to portray) the global similarity-and-dissimilarity structure of some external domain of features, kinds, or properties. Which and whether any of those concepts happens to be regularly *activated*, via the perceptual transduction of some external stimuli, is entirely incidental and inessential to the meaning or semantic content of that concept. To put it bluntly, perceivability is an epistemological matter, not a semantic matter (see Churchland 1979, 13–14).

One can see the mutually orthogonal character of these two important dimensions by considering the semantic consequences, for your assembled *color* concepts, of the sudden onset of total blindness. The semantic consequences are zero. Your capacity for the normal range of visual sensations may thus be lost (your opponent-process system, for example, is now out of business), and your epistemological capacity for the reliable and spontaneous perceptual application of any one of your color concepts may thus have disappeared. But the semantic contents of your several color concepts do not disappear along with those epistemological capacities, as Indicator Semantics entails that they must. Indeed, the semantic contents of those concepts are exactly the same, after the onset of blindness, as they were before. You can still participate, knowledgeably and intelligibly, in conversations concerning the colors of objects. The semantic content of those concepts must therefore be grounded in something other than mere 'reliable indication.'

Surviving *memories* (if any) of past experiences are not what keeps those concepts 'semantically alive' in the people who remember them. One finds the same situation if one considers people who have been color blind, or even totally blind, from birth. As is well known, such people learn to use—though not through observation—the same color vocabulary that the rest of us learn, although they learn it much more slowly. They come to absorb much of the same general knowledge about which colors are characteristic of which kinds of things, and about what causal properties they typically display in consequence. And they come to make, as a matter of course, all or most of the same inferences, involving those terms, as are made by the rest of us. Indeed, if one asks such people to make a series of similarity and dissimilarity judgments in response to questions such as, "As you understand things, is orange more similar to red than it is to blue?," their answers reveal the possession of a structured similarity space with the same gross internal relations displayed among the color concepts of a fully sighted person. Evidently, they acquire, if only roughly, the same abstract conceptual map as the rest of us. What those folks are missing is only our primary (i.e., our perceptual or sensational) *means of activating* the various points within that independently structured conceptual map. Such people are limited to comparatively roundabout *inferences* as a means to activating those points. But the overall family of such points, in them, is a map much like the map possessed by regular-sighted individuals. And as the Domain Portrayal approach insists, it is that shared map that bestows semantic content on the concepts it contains. It plainly and incontestably does so for the congenitally color-blind folks. But that shared map does it for us fully color-*sighted* folks no less than for our color-*blind* brothers and sisters.

I choose the case of our color concepts, to illustrate this general semantic point, only for its vividness (and perhaps for its shock value). In fact, the illustration is easily generalized. Choose any family of familiar observational terms, for whichever sensory modality you like (the family of terms for *temperature*, for example), and ask what happens to the semantic content of those terms, for their users, if every user has the relevant sensory modality suddenly and permanently disabled. The correct answer is that the relevant family of terms, which used to be at least partly *observational* for those users, has now become a family of purely *theoretical* terms. But those terms can still play, and surely will play, much the same descriptive, predictive, explanatory, and manipulative roles that they have always played in the conceptual commerce of those users. For their meanings are alive, and well, *and quite unchanged*, despite the loss of the relevant sensory modality—just as Domain-Portrayal Semantics predicts.

If Indicator Semantics (unlike Domain-Portrayal Semantics) has, as we saw, a secondary problem accounting for the content of our *non*perceptual concepts, it has at least two major problems with its core account of the semantic contents of our allegedly basic perceptual concepts as well. The first problem is *epistemological*, and it arises from the avowedly atomistic nature of the account proposed. Specifically, each perceptual concept, C, is said to acquire or enjoy its semantic content independently of every other concept in its possessor's repertoire, and independently of whatever knowledge or beliefs, involving C, its possessor may or may not have (see Fodor 1990). This semantic atomism is a simple consequence of the presumed fact that whether or not C bears the required lawlike connection, to an external feature, that makes its tokening a reliable indicator of that feature, is a matter that is independent of whatever other concepts its possessor may or may not command, and of whatever beliefs he may or may not embrace.

Such a semantic atomism is entirely plausible, even inescapable, if our prototypical examples of 'reliable indication' are the behaviors of hypersimple energy-transducing measuring instruments such as the thermometer and the voltmeter cited above. For the success of their peculiar feature indications is in no way dependent on any knowledge that they might possess, nor on any auxiliary indicatory talents they might have in addition. In fact, they have neither, nor would anything change if they did.

So far, so good. But a problem begins to emerge, an epistemological problem, when we shift from such cognitively innocent cases of purely transductive indication to the cases of real interest to us—namely, the normal application of our typical observation terms (or, more strictly, the normal activation of our typical observation concepts), such as *face, cookie, doll*, and *sock*. The problem is simple. There *are no* laws of nature that comprehend these things, *qua* faces, cookies, dolls, or socks. All of these features, to be sure, have causal effects on the sensory apparatus of humans, but those effects are diffuse, context-dependent, high-dimensional, and very hard to distinguish, as a class, from the class of perceptual effects that arise from many other things. Think again of the face-recognition network and its acquired capacity for reliably indicating the presence of distinct but highly similar individuals—the presence of Mary, for example, as opposed to Liz. A single *law of nature* for each discriminable individual, or even for each facial type, is simply not to be had.[9]

That is why a student neural network has to struggle so mightily to assemble an even roughly reliable profile of diagnostic dimensions, a profile that will allow it to distinguish *socks*, for example, from such things as shoes, boots, slippers, and sandals, not to mention mittens, gloves,

9. This paragraph, and the several paragraphs following, are drawn from an earlier essay entitled "Neurosemantics: On the Mapping of Minds and the Portrayal of Worlds," originally presented at an interdisciplinary conference on the mind–brain in 2000 in Milan. Figures 2.18 through 2.21 in the next section, concerned with higher-dimensional map congruences, are also drawn from that essay.

tea-cozies, and cloth hand-puppets. All children acquire a reliable indicative response to the presence of socks, but such a response is deeply dependent on that acquired command of a broad range of individually inadequate but overlapping and collectively trustworthy diagnostic dimensions. Those assembled dimensions are what structure the network's activation space, and their acquired sensitivities are what dictate whatever internal similarity metric the matured space displays. Without such a structured family of activation-space categories—that is, without at least some systematic *understanding* of the character of the feature being discriminated, and its manifold relations to other features whose character is also understood—the network will never be able to reliably discriminate that feature from others in its perceptual environment. There simply are no laws of nature *adequate* to make the desired concept-to-world connections directly.

Accordingly, the only access a human child can hope to have to the several features cited above, and to almost every other feature of conceivable interest to it, lies through a high-dimensional informational filter of the well-instructed and well-informed sort displayed in a multilayer device such as Cottrell's mature face network. But such a filter *already* constitutes what atomistic Indicator Semantics portrays as unnecessary—namely, a nontrivial weave of acquired *knowledge* about the various features in question, and indeed, about the entire feature-domain that contains them. We are thus led back, and rather swiftly, to semantic holism.

Before leaving this point, let me emphasize that this is not just another argument for semantic holism. The present argument is aimed squarely at Fodor's atomism in particular, in that the very kinds of causal/informational connections that he deems necessary to meaning are in general *impossible*, save as they are *made* possible by the grace of the accumulated knit of background knowledge deemed essential to meaning by the semantic holist. That alone is what makes subtle, complex, and deeply context-dependent features perceptually discriminable by any cognitive system. Indeed, it is worth suggesting that the selection pressures to produce these ever more penetrating context-dependent discriminative responses to the environment are precisely what drove the evolutionary development of multilayered networks and higher cognitive processes in the first place. Without such well-informed discriminative processes thus lifted into place, we would all be stuck at the cognitive level of the mindless mercury column in the thoughtless thermometer and the uncomprehending needle position in the vacant voltmeter.

Beyond that trivial level, therefore, we should adopt it as a (pre-)Revolutionary Principle that there can be "No Representation without at least

Some Comprehension." And the reason for embracing such a principle is not that we have been talked into some a priori analysis of the notion of "representation." The reason is that, in general, representations cannot do their cognitive *jobs*—such as allowing us to make relevant and reliable discriminative responses to the environment, and subsequently to steer our way through that environment—in an informational vacuum. If you are to have any hope of recognizing and locating your place and your situation within a complex environment, you had better know, beforehand, a good deal about the structure and organization of that environment.

In sum, no cognitive system could ever *possess* the intricate kinds of causal or informational sensitivities variously deemed necessary by atomistic semantic theories, save by virtue of its possessing some systematic grasp of the world's categorical/causal structure. The embedding network of presumptive *general* information so central to semantic holism is not the *post facto* 'luxury' it is on Fodor's approach. It is *epistemologically* essential to any discriminative system above the level of an earthworm.[10] Plato, once again, would approve.

Epistemology aside, Indicator Semantics has a further defect concerning our perceptual concepts, this time with regard to the assignment of semantic content itself. This complaint has been around for at least a quarter-century and has found many independent voices, but it makes even more poignant sense from the perspective of Domain Portrayal Semantics. The objection concerns the endlessly different ways in which distinct perceivers—p_1, p_2, . . . , p_n—might *conceive* of the very same objective feature to which the tokenings of their semantically *diverse* concepts—C_1, C_2, . . . , C_n—all constitute a reliable response or indication. In short, the *same* feature gets indicated in each case, but the indicator-tokens have a *different* semantic content across distinct perceivers. Such cases are not only possible, they are actual; indeed, they are quite common.

"The sky-gods are angry," says the primitive savage, when and only when he sees what we might call a lightning bolt. "The clouds have released some electric fluid," says Benjamin Franklin, when and only when

---

10. After writing this, I learned that Rob Cummins has urged the same general point in his "The Lot of the Causal Theory of Mental Content," *Journal of Philosophy* 94, no. 10 (1997): 535–542. As he puts it there, succinctly, 'distal' properties are 'nontransducible.' I concur. What is novel in the account of this chapter is the specifically connectionist story it provides of how any cognitive creature ever manages to transcend that barrier. But the present account is entirely consistent with Cummins's focal claim that it is our *theories* that put us in contact with 'distal' properties. Indeed, it is the connectionist *version* of Cummins's claim.

he sees exactly what the savage sees. "There is a megajoule flux of charged subatomic particles," says the modern physicist, when and only when she sees the same thing. The concepts activated in these three cases have dramatically *different* meanings or semantic contents, but they are identically caused or activated. Semantic content, it would seem once more, must have its source in something beyond mere reliable indication.

Indeed, the defect here is worse than an unexpected and unwelcome semantic diversity plaguing our observational concepts. Indicator Semantics renders impossible something that is historically quite common, specifically, the chronic or systematic *mis*perception of some aspect of the world. For thousands of years the sincere report "The Sun is rising" has been a reliable indicator of the fact that the moving Earth was rotating the observer into a position where the (unmoving) Sun became visible. But this latter is not what anyone *meant* by such a report, and the official resistance to any such newfangled reinterpretation of the facts was strong enough that people were occasionally burned at the stake for suggesting it. (I here remind the reader of Giordano Bruno and his fatal conflict with the benighted elders of the Roman church.) Clearly, what was reliably *indicated* by tokenings of that expression was one thing; what was standardly *meant*, by normal speakers, was quite another.

Similarly, a perceptual report of the form "There is a star" has been, throughout history, a reliable indicator of the existence and angular location of a gravitationally kindled thermonuclear furnace with a mass over a million times that of the Earth. But that is not what any primitive ever *meant* by the simple expression in quotes. More likely what he meant by "a star" was something closer to "a campfire of the gods," or "a window onto heaven," or "a distant crystalline jewel." Accordingly, and even if one were to concede that Indicator Semantics correctly accounts for the objective *reference* of at least some concepts, it is clear that it fails entirely to capture the *sense, meaning,* or *semantic content* of any concept. Evidently, there is an entire universe of conceptual or semantic diversity to which that approach is simply blind, and on which it has no determinative purchase.

That evident range of semantical possibilities is simply the range of different ways in which distinct individuals, or distinct cultures, can *conceive of* or conceptually *portray* the very same domain of objective phenomena. It is the almost endless range of different ways in which a neuronal activation space can be sculpted, by training, into a coherent family of prototype regions structured by a high-dimensional set of similarity, dissimilarity, inclusion, and exclusion relations. To use yet a third vocabulary (one no doubt familiar to the reader), it is the range of distinct background *theories* with which distinct individuals can differently *interpret* the peripheral sensory flux that all of us humans are fated, by sheer biology, to share.

The metaphor of a map, already deployed in illustration of other aspects of Domain-Portrayal Semantics, handles the intricacies here contemplated quite nicely. For one and the same geographical area can be quite differently portrayed by a variety of distinct maps. The San Diego metropolitan area, for example, receives its most common portrayal in the fold-out street maps available at any service station or AAA office. These maps detail the freeways, surface roads, bridges, and tunnels available to any automobile. But if one is the pilot of a TV news helicopter, one may be more likely to guide one's navigation with a quite different map, one that portrays the landscape's topographical features, the locations and altitudes of the local canyons and mountain peaks, other navigational hazards such as broadcast towers and skyscrapers, and the location of local airstrips with their always busy and always dangerous takeoff and landing paths. Such a map, with its nested isoclines and sundry forbidden zones, will look very different from the street map, although it maps equally well, in its own way, onto the reality below.

Alternatively, if the helicopter belongs to the city's utilities department, a third map might usefully portray the complete grid of the city's storm-drain and sewage-disposal pipes, its fresh-water mains, its natural-gas mains, and its substations and electrical power lines. Or, if one is a conservationist, one might construct a color-coded and texture-coded map of the distributions of the diverse plant and animal species that still live in the canyons, the coastal lagoons, the highlands, and the mesas around which the city has been built. We are now looking at four different maps of the San Diego metropolitan area, each one focused on a different aspect of its objective structure, and each one tuned to the administration of a different set of practical concerns.

Differences in focus aside, maps can also differ substantially in the *accuracy* of their portrayals, and in the detail or *penetration* of the portraits they provide. A San Diego street map might be defective in being fifteen years out of date, wrongly depicting bridges that have since been torn down, and showing empty spaces where now there are new subdivisions. Or its auxiliary, corner-inset map of greater Los Angeles might be a frustration because that much smaller map portrays only the major freeways that run through that city, and none of its myriad surface streets. Finally, of course, the map may be metrically deformed in endless ways.

All of this serves to highlight what Indicator Semantics is inclined to suppress and is ill equipped to explain, namely, that one and the same objective domain can be, and typically will be, *differently conceived or understood* by distinct individuals and cultures, even at the level of our spontaneous perceptual comprehensions. Those differences can be minor, as they probably are between you and me for most domains. But they can also be major, as they frequently are between individuals with very different levels of education and acquired practical skills, and between cultures at different levels of historical and scientific development. Let it be noted, then, that the Domain-Portrayal Semantics held out to us by the sculpted activation-space account of human and animal cognition has no difficulties at all in accounting for the substantial *conceptual diversity* often found across individuals, even at the level of their perceptual concepts, and even if the individuals at issue share identical native sensory equipment. Nor has it any difficulties in characterizing systematic and long-term *mis*conceptions of reality, at both levels of generality, namely, the here-and-now perceptual level and the background time-independent theoretical level. This latter virtue of Domain-Portrayal Semantics is especially important, since fundamental misconceptions of this, that, or the other objective domain are a chronic feature of human intellectual history, and their occasional repair is of central interest to epistemology and the philosophy of science.

At the outset of this section, I remarked that there was something profoundly *right* about the view here under critical (indeed, I hope *withering*) scrutiny. And there is. Specifically, the fleeting states of our perceptual modalities do indeed contain enormous amounts of information, both absolute and conditional, about some aspects of the local environment, no less than do the fleeting states of a measuring instrument such as a thermometer or a voltmeter. But simply containing such information is not enough. For any of that information to become *available* to some cognitive creature or other, the relevant instrument of measurement or detection must be *calibrated* in some fashion or other. To take two familiar examples, we must position a scale of radiating marks behind the needle of the voltmeter, marks labeled '0 *V*,' '5 *V*,' '10 *V*,' and so on. And we must inscribe a series of parallel lines along the length of the thermometer, lines labeled '97°*F*,' '98°*F*,' '99°*F*,' and so forth. Failing such an enduring scale behind or in front of the device's moving indicator, the fleeting states of these instruments will still contain systematic information, but it will not be *available* to anyone but an already omniscient God. Such a background calibrating scale, note well, is the limiting case of a *map*. The polar array of the voltmeter scale constitutes a one-dimensional map, appropriately labeled, of 'voltage space.' Similarly, the set of cross-hatches on the bedside thermometer constitutes a one-dimensional map, appropriately labeled, of a small segment of 'temperature space.' Finally, the moving needle itself and the moving column of mercury each constitute a mobile "You are here" indicator, just as we imagined for the two-dimensional highway map.

Let us return, for a moment, to that GPS-connected highway map—the one with a brilliant red laser-dot flanked by a glowing "You are here" icon that *moves* across the manifold of printed map-elements. Imagine now that those background map-elements are all suddenly *erased*, leaving only the laser's moving indicator-dot, glowing against a now-featureless fold-out white paper background about three feet square. This moving red dot, the 'place indicator' may retain its original GPS connection, and its changing position on the now-blank square may retain exactly the same sorts of momentary geographical information (say, "you are now at the intersection of highways I-5 and 805") that it contained before the erasure of all of the background map-elements. (It contains that information in the sense that the dot would not be exactly *here* on the three-foot square unless the map itself *were* physically placed exactly at the intersection of highways I-5 and 805.) But with the erasure described, the map-with-moving-dot has now been rendered completely useless as a means of determining your geographical location, or as a means of navigating our interstate highways. The moving red dot may *have* the relevant information, but that information is now lost to *us*. For there is no longer a global and systematic background representation of the entire road-system reality, some determinate *part* of which can be fingered by the dot as "where, in the space of acknowledged possible locations, we are currently located."

Indeed, from the point of view of the original (intact) map's user, the closest he ever gets to whatever God's-eye information might be objectively contained in the glowing dot's current position is the much more modest semantic content embodied in the map's own humble and much more limited portrayal of its intended domain. For example, the opaque, unavailable, God's-eye information contained in the dot's current position on the highway map might well be, or might include, "You are exactly on the fault-line where the Pacific tectonic crustal plate meets, and is locked against, the North American tectonic crustal plate." And yet, your laser-enhanced map indicates only that you are in the Santa Cruz mountains south of San Francisco, on Page Mill Road, about three miles west of highway 280; for that map is just a simple highway map, and is wholly innocent of plate tectonics.

For an example closer to home, the God's-eye information objectively contained in your spontaneous perceptual judgment about the temperature in this room might well be (unbeknownst to you) "The mean kinetic energy of the molecules in this room is 6.4×10^{-21} joules" ($\approx 75°F$). But the actual semantic content of your judgment is likely the more humble and mundane "It's warm in here," a content that reflects the peculiar conceptual map that *you* have brought to the business of representing the world. And so it is with all perceptual judgments.

In sum, the objective, God's-eye information that may be invisibly contained in our fleeting cognitive states is not the default *semantic content* of those states, to be exchanged transparently and at face value in the ongoing cognitive commerce of normal human beings. Rather, that presumptive but hidden information is, at best, what we *aspire* to know and appreciate at the Ideal End of our representational efforts. It is the complex and very distant *goal* of our epistemic adventure. It is what we shall achieve when the high-dimensional maps that we construct have finally become perfectly accurate and infinitely detailed. In short, it is information we shall likely never possess. But there is no reason we cannot severally approximate it, ever more closely, and ever more comprehensively, with a series of interim maps of more modest, but still serviceable, success.

This puts us, once more, moderately at odds with Immanuel Kant. For Kant argued that the character and structure of the independent world of things-in-themselves was forever inaccessible to both human experience and human reason. Estimates of how accurately any conceptual map might actually portray the transcendent *dinge-an-sich* are therefore misconceived and permanently out of the question—the idle conceits of someone who has failed to grasp the fundamental Kantian distinction between the empirical and the transcendental points of view.

I announce myself guilty of exactly that failure of appreciation, and of those very same conceits (of which, more anon). But before addressing these conflicts, let me highlight, and commend to your embrace, two *positive* Kantian lessons that emerge quite naturally from the preceding discussion.

You will recall the novel highway map augmented with the GPS-activated 'sensory system' that steered a brilliant indicator-dot around the map's surface. And you may recall that, with the complete erasure of the background map-elements, the semantic significance of the dot's current position evaporated. You may also recall that, without *some* empirical indication of one's actual position on the *original* map (if only your fingertip, guided by your visual reading of the local street signs), that map would

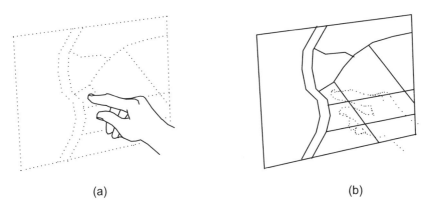

(a)   (b)

Figure 2.16
(a) Intuitions without concepts. (b) Concepts without intuitions.

also have been rendered useless for navigation, despite its objective homomorphism with the target domain.

These two points recall Kant's own observations that intuitions without concepts are empty, while concepts without intuitions are blind. For it is clear from the preceding discussion that indicator-states or locations or configurations that fail to be located against the background of an antecedently structured activation space are uninformative or semantically empty (to anyone except an omniscient God), while structured activation spaces without any trustworthy "You are here" activations within them are epistemologically blind (even if the background space as a whole does correctly represent the global structure of some external domain). More tersely, fingertips without maps are empty; maps without fingertips are blind. See figure 2.16.

The point that indicator-states are semantically empty, save as they appear against the background of an embedding map of some sort, has in one sense been made before. In an earlier exchange with Fodor, I argued that any observation sentence '*Fa*' will be devoid of meaning in the complete absence of some set or other of general background commitments or theory, such as '$(x)(Fx \supset Gx)$' and '$(x)((Fx \ \& \ Hx) \supset \sim Kx)$', held by the perceiver who tokens that sentence (Churchland 1988). The reason is that, in the absence of such an embedding background, the assertion or tokening of '*Fa*' will be *computationally inert*. It will be totally without any distinguishing logical consequences or inferential significance for the rest of the perceiver's cognitive commitments. Accordingly, if that observational tokening is to play any role at all in the perceiver's subsequent cognitive economy, it *must* occur against a background of some such general cognitive commitments already in place. With them in place (see above), *'Fa'* will then imply *'Ga'*; it will be incompatible with *'Ha* & *Ka'*; and so forth. Without them, *'Fa'* will be a gear that engages nothing. And so for any other observation report.

This early expression of the point, however, has the defect of being limited to linguaformal background commitments, and to the computational activities comprehended by classical logic. A more recent expression by Grush (1997), also aimed against Indicator Semantics, repairs that limitation. As he argues the point, an indicator state must be semantically empty save as it possesses some *structure* that plays a determinate computational role in the subsequent *transformations* of that representation into other representations higher up in the information-processing ladder. And without a distinct transformational profile, no representation can have a meaning or semantic content that is distinct from any other representation within that information-processing economy.

To apply this point to feed-forward neural networks in particular, the unique structure of any vectorial representation resides in the constituting pattern of activation levels across the neuronal population involved. That pattern is one of many possible alternative activation-patterns variously located within that activation space, each of which displays a different transformational profile when it gets projected through the matrix of synaptic connections meeting the next rung in the processing ladder. Once again, semantic content is to be found in the well-structured range of alternative representational points within a background activation space, and diverse semantic contents will show themselves in the well-structured range of alternative computational consequences that they have for the cognitive system in which they are embedded. Same semantic contents: same computational consequences.

9 On the Identity/Similarity of Conceptual Frameworks across Distinct Individuals

The view on the table is that to possess a conceptual framework for any given feature-domain is to possess a high-dimensional activation space that has been metrically sculpted into a maplike internal structure. The representational significance or semantic content of any given map-element therein (e.g., a specific metrically compacted prototype region) is determined by the unique profile of its many proximity and distance relations to all of the *other* map-elements within that same space. This semantic holism is entirely appropriate, because it is only when we draw back a step, and focus on the conceptual framework or structured activation space *taken as a whole*, that we can see what needs to be seen, namely, the global homomorphism between its acquired internal structure on the one hand, and the independent structure of the external feature-space or property-domain that it more or less successfully portrays, on the other. Specifically, proximity and distality relations within the former (inner) space correspond to similarity and difference relations within the latter (outer) space. The former is a neuronal- activation space, and the latter is an objective-feature space.

If this holistic and brain-based semantic theory is to claim a decisive superiority over the atomistic alternatives of the preceding two sections, it must pass a fundamental and very demanding test that the alternatives both fail. In particular, it must provide an adequate criterion of *identity* and/or *similarity* of meaning or semantic content across distinct cognitive individuals, and across the same individual considered at different stages of his or her cognitive development. Some theorists (Fodor and Lepore 1992) have argued at length that a 'state-space semantics' of the sort here at issue is doomed to failure on precisely this score. As you are about to see, however, its ability to meet this requirement is one of the most effortless and obvious *virtues* of the state-space or domain-portrayal semantics here under development.

We begin, as before, with the metaphor of a map—or, in this case, a pair of maps. The question before us is: Can we make useful and systematic sense of the notion of *sameness* across distinct maps? Let us begin with the easiest case—two identical maps—and then move outward from there. Consider the upper two road maps of figure 2.17 (plate 9). Map *a* and map *b* are both entirely free of place-names, or any other giveaway semantic labels, and both are presented against a background that is free of any orienting or locating grid, such as lines of latitude and longitude.[11]

Any map will contain a finite number of *map elements*, the nature of which will depend on the sort of map involved. They might be small, closed curves, as within a topographic map, for representing sundry mountaintops. They might be metrically compacted regions, as with the

11. Here again, I draw (at rather greater length) on an example cited in an earlier paper, "What Happens to Reliabilism When It Is Liberated from the Propositional Attitudes?" originally presented at a conference in honor of the philosophy of Alvin Goldman at Tucson, Arizona, in 2000.

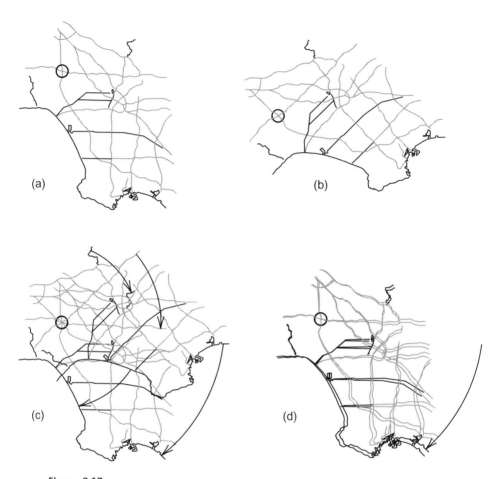

Figure 2.17
Two distinct maps superimposed and rotated to find a systematic homomorphism.
See plate 9.

prototype points within a sculpted activation space. Or they might be intersecting wiggly lines, as in a map of a modern city's freeway system.

For purposes of illustration, let us take this last case as representative of the larger population. In particular, consider the respective sets of freeway *intersections* found on maps (a) and (b) in figure 2.17. There are finitely many such map elements in each set. At random, choose one such element from map (a)—say, the intersection marked with a circle close to the top—and one such element from map (b)—say, the intersection marked with a circle toward the left.

Now superimpose map (b) over map (a) so that the two map elements circled above coincide exactly, as portrayed in map (c) in figure 2.17. (Suppose that both maps are printed on transparent sheets, so that this superposition is visible.) As you may observe in map (c), this yields a chaotic tangle of mismatched lines and intersections. But now slowly *rotate* the upper map around the chosen intersection(s), while holding the lower map motionless. Within something less than 360°, you will observe all of the elements in the upper map collectively converge on the corresponding elements of the lower map, as penultimately portrayed in map (d) in figure 2.17. Map (b) simply 'disappears,' as it were, into map (a). The two maps are thus revealed to be qualitatively identical, in that there is a one-to-one correspondence or mapping between the elements of each map, a mapping that *preserves all of the distance relations* within each. They are, in a word, isomorphic. In point of fact, they are both same-scale maps of the major freeways, surface streets, and coastline of the city of Los Angeles. Map (b) is simply rotated some 30° counterclockwise, in its initial position on the printed page, relative to map (a).

Of course, I deliberately narrowed your search space by choosing the specific pair of circled intersections at issue. But a finite search of all possible intersection pairs would have found that pair sooner or later (or any one of some two dozen other intersection pairs, around which a comparable rotation would yield a global convergence just as reliably). We thus have an effective procedure for determining, of two unlabeled, unsituated, and causally anonymous maps, whether they embody the *same portrayal* of whatever it is that they purport to portray. If they do indeed embody the same portrayal, then, for some superposition of respective map elements, the across-map distances, between any distinct map element in (a) and its 'nearest distinct map element' in (b), will fall collectively to zero for some rotation of map (b) around some appropriate superposition point.

Even differences in scale, between the two maps, need not defeat us. For we can simply repeat the procedure described above with a series of distinct magnifications or minifications of map (b). As the two maps *approach* being of the same scale, so will the across-map distances just mentioned collectively approach the limit of zero, for some superposition and mutual rotation.

The procedure described is also robust for finding *partial* identities, as when map (a) contains everything in map (b), but a good deal more detail—for example, detail concerning the many minor surface streets of Los Angeles. In that case, some superposition of intersections and some rotation will again make map (b) 'disappear into' map (a). But in that case, the relation-preserving mapping takes the set of map elements of (b) *into* the set of map elements of (a), rather than *onto* those elements. The mapping finds a homomorphism rather than a perfect (i.e., mutual) isomorphism.

The procedure is robust once more for maps that *partially overlap* one another, as when map (a) includes not only the freeways of central Los Angeles, but also those of Burbank to the north, while map (b) includes not only those same freeways of central Los Angeles, but also those of Long Beach to the south. Here a *part* of map (a) can be made to 'disappear into' a *part* of map (b), namely, the respective parts that happen to represent central Los Angeles. The other parts are simply left out of this partial, but still striking and a priori improbable, relation-preserving mapping.

Finally, and most importantly, the procedure is robust for maps of any dimensionality whatsoever, and not just for our illustrative case of two-dimensional highway maps. I hope the reader here will forgive my brief reuse of an agreeably transparent example that receives much more detailed discussion in an earlier paper (Churchland 2001) written in response to Fodor and Lepore (1999), and Evan Tiffany (1999). You may observe this homomorphism-finding procedure at work with a pair of simple *three-*dimensional maps, each with exactly four floating 'intersections' joined by six straight-line 'highways,' forming a pair of irregular tetrahedrons, as shown in figure 2.18. To minimize our search space, superimpose the longest edge of tetrahedron (b) onto the longest edge of tetrahedron (a), bringing the rest of the tetrahedron along for the ride.

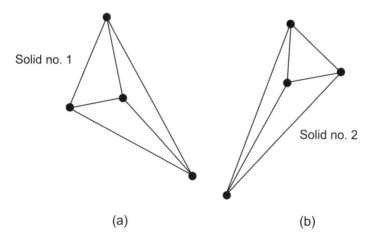

Solid no. 1

Solid no. 2

(a)   (b)

Figure 2.18
Two irregular tetrahedrons.

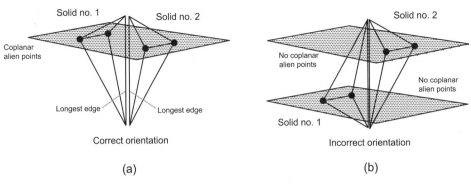

Figure 2.19
Finding a mutual mapping: first steps.

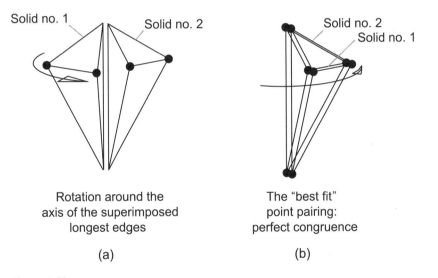

Figure 2.20
Finding a mutual mapping: final steps.

There are two possible orientations here, as shown in figure 2.19. Choose whichever orientation finds the largest number of *alien* coplanar map elements, as judged relative to planes orthogonal to the two superimposed edges. Now rotate tetrahedron (b) around that superimposed axis until it 'disappears into' tetrahedron (a), as shown in figure 2.20. Once again, we have found a one-to-one mapping of the elements of map (b) onto map (a), a mapping that preserves all of the distance relations within each map. In that sense, the two maps are qualitatively identical, despite their having different initial orientations, within objective three-dimensional space, in fig. 2.18. Note in particular that (1) the procedure described *finds* the relevant structure-to-structure mapping without the benefit of any semantic labels or information about special causal connections with any features in the external world, and that (2) the sheer *existence* of that one-to-one mapping has nothing to with the existence or nonexistence of any such extraneous information. The notion of "sameness of meaning" here on display is a strictly *internalist* notion. Roughly speaking, it is the map-analogue of the notion of "narrow content" familiar to us from the recent philosophical literature, or the map-analogue of Frege's notion of *Sinn* (sense) as opposed to his notion of *Bedeutung* (reference).

It is also something else. It is exactly the notion of sameness and difference that we need to make sense of the semantic significance of the various structures that emerge, during training, within the activation spaces of large populations of neurons, both biological and artificial. That those structures are best seen as high-dimensional *maps* of various objective *feature spaces* is one of the major themes of this volume, a theme that, I own once more, is not yet firmly established. But note well that, *if* those structures *are* maps of the kind at issue, then our semantic worries are over. Or, more modestly, *if* those structures are as here described, then we have a well-behaved and systematic criterion for samenesses-of-portrayal and differences-of-portrayal across the representational systems of distinct neural networks.

Moreover, it is a criterion that fits hand-in-glove with a well-behaved and systematic account of how such portrayals can both succeed and fail in the business of representing aspects of the external world. The identity criterion and the representational account are the same ones that serve us so well in the familiar case of maps, as now more broadly conceived to include 'maps' of any dimensionality, 'maps' that address abstract *feature spaces* as well as concrete geographical spaces. What has eluded us for so many decades—indeed, for many centuries—is how the *brain* can be the seat of something on the order of at least a thousand distinct high-dimensional maps, all of them interacting with one another, with the external world, and with the body's motor system, on a time scale of milliseconds. But as the scientific picture of our own neuroanatomical structure and neurophysiological activity slowly converges to a sharp focus, this is indeed how things appear to be.

For many reasons, a genuinely perfect identity of conceptual frameworks across any two individual brains is highly unlikely. The sort of case just discussed is only the ideal, which real cases tend, at best, to

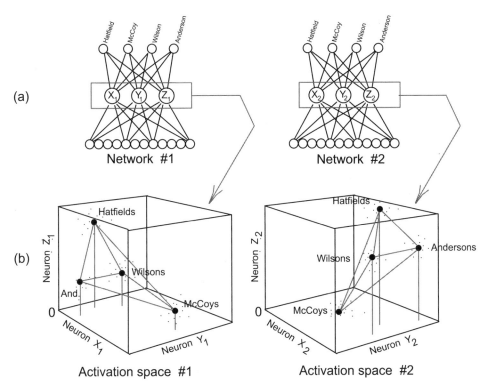

Figure 2.21
Two face maps with different activation-space orientations.

approximate. We need, therefore, a numerical measure of conceptual *similarity*, short of strict *identity*, if we are to assess real cases in an objective manner. Such a measure has already been articulated. Some readers will recognize the two irregular tetrahedrons, of figure 2.18 above, from an earlier (2001) paper on that topic (see Churchland 1998). Originally, those tetrahedrons represented the respective prototype families learned by two face-recognition networks, as shown in figure 2.21.

These two sculpted activation spaces, one for each of the two middle-rung neuronal populations of the two networks, are (fictional) maps of a range of possible human faces drawn from the rural mountain communities of the historical American Southeast. The networks have been trained to discriminate, let us suppose, membership in one of four possible extended families—the Hatfields, the McCoys, the Wilsons, and the Andersons, as pictured in figure 2.21. That training yielded, for the middle-rung space of each network, four metrically compacted prototype regions (marked by the four black dots) separated by six distances (marked by the six straight lines). That is, the training produced two irregular but tetrahedron-shaped three-dimensional maps of the peculiar range of faces (namely, the members of the four extended families) contained in its training set.[12]

As you can see, the two quartets of prototype points are very differently oriented and located within their respective activation spaces—that is, relative to the basic neuronal *axes* of the two spaces. But the families of *distance relations* that configure or mutually situate the respective prototype points in each quartet are identical across the two activation spaces. That is to say, the two networks have settled into the *same* portrayal of the objective similarity and distance relations at issue, the relations that unite and divide the various faces in their common training set. The structure-finding procedure described above will therefore find a perfect isomorphism between them.

But now suppose that those two tetrahedrons are *not* perfectly identical, but only roughly *similar*, an outcome entirely likely for real networks undergoing real training. This poses no problem. The procedure at issue will still find the 'best possible' superposition of these two maps, even if no perfect superposition exists. The principal difference in these imperfect cases is that the mutual rotation of the two maps eventually finds an angle at which, for all of the black-dot map elements, the sum of all of the 'distance-to-the-nearest-alien-map-element' increments falls not to zero, but to some *minimum*, relative to all possible edge-pairings and mutual rotations. Thus, rough correspondences, if they exist, can always be identified, in addition to the perfect ones.

The similarity measure needed here is agreeably simple. First, find the best available superposition of the two structures to be compared. Then, for any two corresponding edges, L_1 and L_2 (one from each of the two structures being compared), divide the absolute difference in length between them by the sum of those two lengths. This will always yield a fraction between 0 and 1, tending toward 0 as the two lengths approach identity, and tending toward 1 as the two lengths diverge. Now take the *average* of all such fractions, as computed across each pair of corresponding edges, and subtract it from 1. That yields the desired similarity measure.

12. This is once again a deliberate cartoon. A mere three dimensions will not sustain a very penetrating map. Recall that Cottrell's network required 80 dimensions to sustain its modest grip on facial-feature space. But visual access, to the translational points here at issue, once more requires a low-dimensional fiction. Hence, the entirely fictional network of figure 2.21.

It, too, will always be a number between 1 and 0, where 1 represents perfect similarity and 0 represents no relevant similarity at all. Canonically,

$$Sim = 1 - avg. \ [\ (| \ L_1 - L_2 \ |) \div (L_1 + L_2) \ ].$$

For the two deliberately identical tetrahedrons pictured above, the difference between any two corresponding edges, L_1 and L_2, is always 0; and so the quotient of the fraction within the square brackets is always 0; and so the average of all such fractions is also 0; and so Sim = 1 - 0, or exactly 1, as desired. For tetrahedrons of progressively less perfect geometrical similarity, the value for Sim will be progressively smaller than 1, tending toward 0 as their respective shapes diverge ever more widely, as when, for example, one of the two tetrahedrons becomes progressively longer, narrower, and more needle-like.

Note that this particular similarity measure is well-defined only for that portion of the respective portrayals that is *shared* by the two maps at issue. As we noted above—concerning a map for Los Angeles and Burbank to the north, and another map for Los Angeles and Long Beach to the south—two maps may overlap nicely for two-thirds of their respective extents, and yet each has a distinct structure remaining, a structure differently connected to their shared overlap. A fractional measure of such overlap is always possible, as when 47 percent of map (a) overlaps 13 percent of some rather larger map (b). This *overlap* measure is distinct from the similarity-of-*structure* measure, Sim, defined above, but it is still highly relevant to our interests in finding the occasional 'translational equivalence' across distinct maps.

So is a third measure, having to do with comparative *detail*, as when the entire extent of a very *sketchy* map (a) 'disappears into' the entire extent of a much more detailed map (b). Here again a fractional measure is obvious. Both maps have a finite number of discrete map elements. If map (a)'s elements correspond to a subset of map (b)'s elements—20 percent of them, let us suppose—then map (a) is only one-fifth as detailed as map (b). As well, two maps may superimpose perfectly in their spatial extents, and in 80 percent of their respective map elements, and yet each may contain a residual set of *un*paired map elements variously interpolated among the elements in their shared 80 percent. In such a case, the two maps are equally detailed, but they are 20 percent concerned with different *kinds* of details.

Evidently, map pairs can display a variety of partial and/or proximate homomorphisms, all of which are interesting, but all of which are familiar—and unproblematic—from the case of two-dimensional road maps.

Nothing essential changes as the dimensionality of the maps involved climbs into the hundreds, thousands, and millions of dimensions. And nothing essential changes when a *position* within a map is registered by an *n*-tuple of possible simultaneous activation levels $<a_1, a_2, a_3, \ldots, a_n>$ within a population of n neurons, instead of by a possible ordered pair of coordinates $<x, y>$ on a two-dimensional paper surface. We can still speak, with a clear and organized conscience, of the sundry structures that may exist within such spaces, and of the various forms of partial and proximate homomorphisms that may obtain between such structures.

Indeed, we already do speak of them, though without full appreciation of what we are doing. A young child's conception of the domain of possible *pets* is quite properly said to be a simplified or less-detailed analogue of an adult's conception of the same domain. For the child, the central categories of *dog*, *cat*, and *bird*, though mutually configured in roughly the same fashion as in an adult, do not yet divide into a systematic set of subordinate categories such as *poodle*, *spaniel*, and *lab*, and so on within the superordinate category *dog*; nor into such subordinate categories as *tabby*, *Siamese*, and *Persian*, and so on within the superordinate category *cat*; nor into *canary*, *budgie*, and *parrot*, and so on within the superordinate category *bird*. Although the child's overall conception of the domain of pets may, for some suitable superposition, 'rotate without remainder' into the adult's conception, the latter is plainly a more detailed portrayal of the internal structure of that target domain.

It may also be 'larger,' in the fashion of the Los-Angeles-plus-Burbank map over the Los Angeles map. The adult will likely appreciate, for example, that monkeys, hamsters, gerbils, iguanas, and even snakes regularly serve as pets for some people, while the possibility of such an odd arrangement has not even yet occurred to the child. Similarly, the child's conception of the domain of pets may outstrip the adult's in some small way, as when the child keeps a humble sow-bug in a matchbox for a week or two, faithfully feeding it and talking to it, a 'social' arrangement whose eventual discovery produces abject amazement in his adult parents.

Less humdrum examples of such partial overlap of global conceptual frameworks would include the three neighborhood acquaintances who deploy the same family of concepts while shopping at the local supermarket, but deploy very different concepts when they go to work. The first shopper does astrophysics and extragalactic astronomy. The second uses PCR (polymerase chain reaction) techniques to help identify criminals in a forensic police lab. And the third does economic analysis to guide large-scale trading on the NYSE. They may converse freely with one another on domestic topics, while standing in line at the checkout counter. But if the topic turns to the intricacies of astronomy (or biochemistry, or economic policy), they will probably have difficulty even speaking with one another.

Common politeness then demands (if any useful conversation is to take place at all) that the astronomer, for example, somehow identify the sparse *sub*structure of his own astronomical conceptions that *is* possessed by his interlocutors, and then pitch his astronomical conversation at a level appropriate to their more elementary understandings of that domain. People do this sort of thing all the time, most often with children, but quite regularly with other adults as well. Evidently, recognizing and dealing with such conceptual disparities, as were described above in explicitly geometrical terms, are recognitional and practical skills that ordinary people already possess. And for good reason: such conceptual disparities are extremely common.[13]

Our day-to-day grip on such conceptual similarities and disparities is typically grounded in the evident similarities and disparities in (1) the overtly linguistic vocabularies commanded by the two thinkers involved, and/or (2) the contents and depth of their general declarative knowledge involving that vocabulary. But such a linguaformal grip remains a very superficial grip for even the best cases, and it provides no grip at all for those many cases (recall the artificial crab's sensorimotor coordination) where our acquired understanding constitutes or includes an inarticulable *skill* of some kind. A lexicon of general terms and a body of accepted sentences containing them will indirectly reflect *some* of the acquired structure of *some* neuronal activation spaces, but it will provide a low-dimensional projection of the high-dimensional reality even there, and no relevant projection at all in many others.

To get a direct, comprehensive, and universal grip on the internal structure of any given neuronal activation space will require that we get much closer to the computational machinery at work in such cases. Most

13. The notion of partially overlapping maps thus provides a natural and entirely realistic solution to the problem, urged by Fodor and Lepore, and more recently by F. C. Garzon (2000), of the inevitable diversity of collateral information across distinct cognitive agents. Such diversity indeed entails that such cognitive agents will have different cognitive maps, but those different maps may still contain overlapping or homomorphic *sub*structures that correspond, well or badly, to the same external feature-domain. And as in the case of the overlapping street-maps discussed above, both the extent of the overlap and the degree of the homomorphism can be objectively measured.

obviously, we might try to specify the connection strengths or 'weights' for the entire matrix of synaptic connections meeting that neuronal population. It is the acquired configuration of these connection weights, recall, that effects the transformation of any input vector into a unique output vector. And so it is that acquired configuration of connection weights that dictates or determines the acquired global structure of the output activation space. Specify the connection weights, and you have uniquely specified the peculiar partitional or maplike structure of the activation space to which those weights collectively give life.

Alternatively, and much more easily, we can sequentially present, to the neuronal population at issue, a systematic *sampling* of the space of possible *input* vectors (that is, the activation patterns across the assembled axons that project to the neuronal population at issue), and then sequentially record the 'daughter' activation patterns thus produced in the target neuronal population. (This is what we did in our opening example, as portrayed in fig. 2.3a.) This procedure will slowly sketch out for us that space's acquired metrical structure, even though we remain ignorant of the actual values of the coefficients of the connection matrix, the matrix that produces the transformations to which we are thus made witness.

With artificial neural networks, such determinations of actual weights and activation vectors are easily made, especially if the structure and the dynamics of the network are modeled within a conventional computer. The many values that make up the current synaptic connection matrix of the model network can be read out on command, as can the input activation vector currently arriving at that matrix, and also the resulting activation vector across the target population of modeled neurons. With these details made transparent to us, the business of determining the identity, the degree of similarity, or the outright incommensurability of the respective conceptual structures embodied in two distinct artificial neural networks becomes child's play. We need only deploy the procedures and measures discussed earlier in this section, or one of a variety of related procedures that do essentially the same job (see Churchland 1998; Cottrell and Laakso 2000).

In principle, those same procedures and measures apply equally well when the networks at issue are living, biological systems. But in practice, getting the information required to apply them is extraordinarily difficult. The axons in the human visual system—projecting to the LGN from the primary visual cortex, for example—make something in the neighborhood of 10^{12} synaptic connections with their target neurons. Each connection is typically less than a micron across. And tracking the diverse axonal *origins* of the many synapses onto any given neuron, within the mare's nest of interwoven axonal end-branches, is the neuronal analogue of trying to untie Alexander's Gordian knot. A technology that will recover for us the entire connection matrix onto any living neuronal population, even post mortem, is a technology that remains to be invented.

The situation improves slightly if we shift our attention from determining the enduring connection matrix to determining the ephemeral activation vectors that get produced across the receiving population of neurons. We can easily record, in real time, the activation levels of a single neuron, by mechanically inserting a macroscopic, electrolyte-filled glass tube (with a drawn-out microscopic tip) through its cell wall. And with some experimental heroics, we can even record from a dozen neurons or so, simultaneously. But we must remember that with vector coding (that is, population coding), as opposed to localist coding, determining the 'receptive fields' or the 'preferred stimuli' of *individual* neurons tells us next to nothing about the representational activities and insights of the collective population functioning as a whole. As we saw in figure 2.13, concerning the face network, such isolated information may reveal the gross 'subject matter' of that rung, but little else. To get at what we want, namely, the peculiar conceptual structures embodied in that space, we need to track that population's behavior as a collective body.

But tracking the entire population of neurons in any brain area is an endeavor defeated once again by the large numbers—perhaps as many as 10^9—and the intimate proximity of the cells at issue. One cannot insert 10^9 macroscopic glass tubes, simultaneously, into the visual cortex, even if they have been carefully drawn into microscopic tips. There isn't remotely enough room. Packed vertically and in parallel, for maximum efficiency, an array of 10^9 drinking-straw-sized tubes would occupy some 4,000 square meters, an area roughly half the size of a soccer field, all in aid of probing a cortical area the size of a child's mitten.

Determining the entire activation pattern across a typical neuronal population thus appears almost as impossible, with current technology, as recovering the connection matrix that produces it. We are limited, with microscopic physical probes, to sampling a small subspace—perhaps a hundred dimensions or so—of the much higher-dimensional activation space embodied in any real brain area. Such comparatively low-dimensional samplings can still be informative, to be sure. Observed over time, they can reveal, perhaps, the gross categorical structure of the target activation space, though at very poor resolution.[14] And in very tiny neuronal populations (≈ 1,000 neurons or so, as can be found in an insect's nervous system), this resolution gap, between the monitored sample and the entire target population, will be much reduced.

A recently developed optical technique offers a rather better approach to divining the global pattern of simultaneous activations across an entire neuronal population, at least for very simple creatures such as a fly. It exploits the intracellular calcium sensitivity of a specific fluorescent dye. Using a technique called two-photon microscopy, the changing calcium levels (which reflect the changing levels of neuronal excitation) within a tiny cluster of same-sense neurons (called *glomeruli*, on the surface of the olfactory bulb of the fly's brain) can be measured by the levels of local *optical* fluorescence. A false-color map (see fig. 2.22, plate 10) of the relative *levels* of activation across that two-dimensional population of glomeruli can thus be made visually apparent, through a microscope, for any odorant presented to the fly's olfactory antennae.

This technique, as used in this particular example, had not yet reached a spatial resolution that would reveal a single neuron's activation level, but it fell short by only a step or two. And given that the neurons in a single glomerulus all share the *same* topical sensitivity, the false-color map just described *is* a give-away map of the fly's primary olfactory space. As the false-color photos attest, distinct external odorants are coded with distinct patterns across the surface, and the same odorant is coded with the same pattern, even across distinct individual flies.[15]

A second and even more recent example of the same technique concerns the activation-levels of the motor neurons in the motor ganglia of

14. "Resolution," as used here, is a fraction between 0 and 1. It is equal to the number of neurons (i.e., dimensions) actually monitored by the technology at issue, divided by the total number of neurons (i.e., dimensions) in the population (i.e., in the global activation space) at issue. Perfect resolution = 1.

15. Note that the 'olfactory map' here divined is *not* two-dimensional, as the character of the olfactory bulb's two-dimensional surface might suggest. If it were, the fly's current position in olfactory space would be coded by the activation of a single *point* on that biological surface. Instead, current olfactory position is coded, as illustrated, by a complex and extended pattern of activation levels across the entire population of glomeruli. The map here divined is in fact an *n*-dimensional space, where *n* = the number of distinct glomeruli in the olfactory bulb, a number in the hundreds. A specific false-color pattern seen through the microscope thus reflects a specific point within a map-space of hundreds of dimensions. There is indeed a map embodied in the fly's olfactory bulb, but we are seeing it only indirectly in the photos of fig. 2.22.

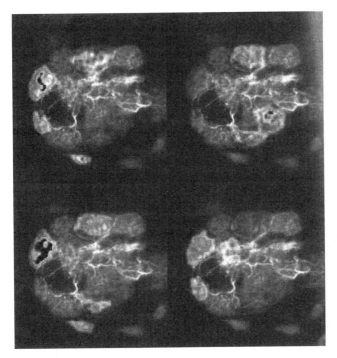

Figure 2.22
Distinct activation patterns in the fly's olfactory glomeruli. (Courtesy of J. W. Wang, Center for Neurobiology and Behavior, Columbia University, New York, USA.) See plate 10.

the leech (see fig. 2.23). Here we can watch, and even film, the changing activation levels of each one of the many neurons in a cluster of twenty or thirty of them, while the leech's nervous system is engaged in the cyclical electrical activity that produces its typical periodic swimming behavior. In this case, remarkably, the individual neurons can be seen to flash on and off, variously in and out of phase with each other, as time unfolds (see Briggman and Kristan 2006).

With advanced techniques such as these, we can at least *test* our hypotheses concerning the computational and representational activities of biological networks, even for awake, behaving creatures. But evidently it will be the very simplest of creatures that yield up their secrets first. The larger brains of mammals, primates, and humans, the ones that interest us the most, will become transparent only very slowly. In the meantime, we can at least *speak* coherently about (1) what manifestly *is* going on inside our transparent *artificial* networks, and about (2) what *might well be*

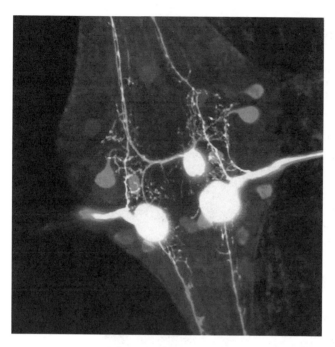

Figure 2.23
Motor neuron activity in the leech. (Courtesy of W. Kristin, Department of Neuroscience, UCSD, La Jolla, CA, USA.) See plate 11.

going on inside their comparatively opaque, but still penetrable, biological counterparts. Even with present technology, this will allow us to pursue the long-term strategy of critically testing our ever-more-sophisticated artificial models against the stubborn but ever-more-transparent biological reality.

In summary, we now command a well-defined *conception* of the identity, the rough similarity, and even the incommensurability of distinct conceptual structures across numerically distinct neuronal populations. We also command—in principle, if not in fact—an effective procedure for *finding* such identities and/or similarities across distinct populations, and a well-behaved measure for *quantifying* their degrees of mutual similarity in cases that fall short of perfect identity. We can also speak sensibly and concretely about the degrees of overlap that distinct frameworks may share, and about the relative detail displayed in their respective portrayals of the world. Thus we behold some additional dividends, this time in the domain of semantic comparison, of adopting the biologically inspired account of both representations-of-the-ephemeral (i.e., activation vectors) and representations-of-the-enduring (i.e., sculpted activation spaces) outlined in the opening chapter. We may add these new benefits to the list of dividends already explored: the Synaptic-Change Model of basic-level learning in the brain, the Map-Indexing Model of Perception, the Vector-Transformation Model of both categorical perception and sensorimotor coordination, and the Vector-Completion Account of the real-time, globally sensitive abductive talents inevitably displayed by any trained network. As well, there are more dividends to come.

3 First-Level Learning, Part 2: On the Evaluation of Maps and Their Generation by Hebbian Learning

1 On the Evaluation of Conceptual Frameworks: A First Pass

If a conceptual framework is a species of *map*—specifically, a high-dimensional structural homomorph of the objective similarity-and-difference structure of some objective, abstract feature-space—then presumably its virtues and vices, and the procedures for determining them, will parallel the comparatively prosaic case of a geographical map. And so, I suggest, they do. The most obvious virtue any map can have is the accuracy of its internal portrayal of the external structures that it purports to, or is deployed to, represent. Here we presume to compare maps not with one another, as in the preceding section, but with the enduring features of the external world.

Alas, this is not as easy as it might seem, even in the case of two-dimensional geographical maps. If Lewis and Clark, just prior to their northwestern voyage of exploration and discovery in 1804, had been handed an exquisitely detailed and faithful paper map of every mountain, valley, river, and creek in the entire continental United States, they would have been in almost no position to *evaluate* that Grand Map's accuracy. For they had no *independent access* to the geographical structures at issue—no space shuttle to lift them into position for a panoramic snapshot of the Pacific Northwest, no high-resolution continental photos to be downloaded from a geostationary spy satellite, nor any Divine Map bequeathed them by a necessarily existent and nondeceiving god.[1]

Or, I should say, they had *almost* no independent access. These gentlemen did have, of course, some modestly detailed land maps of the eastern third of the country, and some rather ragged maritime maps—deformed in longitude, if not in latitude—of the western coastline. These prior maps

1. My apologies to Descartes.

could each be compared with the corresponding parts of the imagined Grand Map whose global virtues are here put up for vindication. An evaluation of at least some largish parts of that Grand Map—its eastern and western extremes, at least—can thereby be undertaken. Simply correct for scale, and then see how closely the various maps overlap.

But we are back, once again, to comparing maps with other maps, maps already validated by some antecedent procedure or other. Even if the shuttle snapshot, the satellite spy photo, and Descartes' Divine Map were made available to our explorers, they are all just more *maps*, after all, despite their variously 'transcendent' genetic credentials. How, we want to know, can any map be evaluated and validated independently of the presumed validity of some *other* map?

A plausible answer suggests that we put the map to work in the business of actually *navigating* the objective domain it presumptively or allegedly portrays. A map tells you *what* to expect and *where* to expect it. If you can somehow keep running track of your unfolding position on any map, you can therefore *compare* the map's unfolding predictions (for Lewis and Clark, their Grand Map predicts a mountain range to the west, an eastward-flowing river descending therefrom, and a narrow westward-leading pass near its headwaters) with your unfolding experience of the local environs. That experience can confirm or disconfirm various aspects of the Grand Map's global portrayal of the domain at issue. Before actually setting out from St. Louis, roughly westward up the Missouri River, Lewis and Clark might not have been able to validate more than a smallish part of the Grand Map imagined above. But had they taken that (highly accurate) map with them on their famous trek, their subsequent adventures would have validated its every detail, at least concerning the geographical path they actually traversed.

There is something obviously right about this answer, but also something problematic. We can begin to see what is suspicious by returning to the case of the automobile with the GPS-activated road map on the dashboard. That computer-stored map, let us assume, is an entirely accurate portrayal of the freeways, surface streets, parks, rivers, lagoons, and coastline of greater San Diego. And an arbitrary tour of the area, guided by that stored map and moving car-icon, will validate all of the map-elements deployed, just as we imagined with our augmented Lewis and Clark. But as did Lewis and Clark during their journey, we will be using *our* native perceptual mechanisms to continuously scout the local environment for features in agreement or disagreement with the stored map.

Insofar, we are introducing one or more *dei ex machina* to fill an obvious gap in our story of how brains work. For unlike the humanly steered automobile with its computer-stored map, the brain does not contain a 'driver' or homunculus with independent perceptual capacities fit for the ongoing evaluation of the accuracy of the brain's stored maps. The brain has to perform such evaluations exclusively with its own resources. And those resources, on our story, are always limited to a substantial population of *acquired maps*, of diverse aspects of the brain-independent world, maps with an ongoing flux of *activated points* somewhere or other within them. The brain is thus inextricably fated to evaluate any one of its acquired maps by comparing the sundry expectations it yields against the ongoing ephemeral deliverances of its many *other* activated internal maps.

One can appreciate the brain's epistemological position by considering a very different arrangement for our GPS-equipped automobile. Let us remove the human driver—the 'homonculus'—entirely. Let us put the car's steering mechanisms under the direct control of the computer that stores the map-plus-moving-icon. Let us program that computer to pursue a hierarchy of goals (e.g., move forward, but stay on the right-hand side of the road, stop at all red lights, avoid collisions with anything). And let us provide with it several *other* cognitive maps with ongoing activations within them also, such as a sensor for the car's distance to other cars, a sensor for red and green traffic lights, a sensor to read street signs, and sensors for acceleration, velocity, and distance traveled. Now we have a completely autonomous vehicle that might navigate the San Diego area quite successfully, at least if its stored map is accurate. If it is not, the system's accelerometers will notify it instantly of any collisions.

We, too, are autonomous vehicles, groping our way forward in a world we only partly understand. The quality of our unfolding navigation is one measure of the accuracy of our acquired conceptual maps, but the quality of our navigation is itself something that must be registered on other maps of antecedently presumed integrity. Some of them, of course, are innate, if rather crude. We have native systems for registering a variety of punishments and rewards. Bodily damage, oxygen deprivation, toxic ingestions, food deprivation, and extremes of heat and cold are all registered in familiar and unpleasant ways. Similarly, local warmth, a balanced metabolic state, a sated appetite, and positive social interactions are all registered, but with an opposite evaluative polarity. Collectively, such registrations exert an extinguishing, or a reinforcing, effect on a variety of navigational behaviors, and ultimately, on the pliable structures of the conceptual maps that steer those behaviors.

But let us remind ourselves that nothing guarantees the accuracy of these native evaluative maps either. They are part of an admirable system for regulating the integrity of our complex internal biochemical milieu—a system of neuronal registration and control presumably shaped by many hundreds of millions of years of evolutionary pressures. But they, too, are imperfect in their structural portrayals of biochemical reality, and they too admit of the occasional false activations, such as auto-immune reactions, even where their background structural portrayals are exemplary. The point of the preceding paragraph is not that the brain's map-evaluating capacities are ultimately grounded in a privileged handful of pleasure and pain maps. Such maps do exist, they do play an evaluative role, and they do steer learning. But they are just a small part of a much larger population of conceptual maps—"in the thousands" is our running guess—*all* of which play a role in the ceaseless business of evaluating both our ephemeral and our enduring representations.

The fact is, many of the maps within the human brain *overlap* each other in the domains that they portray, often substantially. My visual, tactile, and auditory systems, for example, are all in the business of representing the local configuration and character of middle-sized physical objects and processes. I can see the keyboard on which my fingers are currently typing. I can feel the cupped surface of each key as I press it. And I can hear the faint click of each key as it hits bottom. Lifting my eyes to look out the open balcony door, I can see the Eucalyptus trees swaying in the hot Santa Ana wind, hear their dry leaves rustling, and feel the occasional draft of wind cresting through my open balcony door. With such simultaneous and commonly caused activations occurring across a variety of modally distinct but topic-overlapping maps, the brain is in a position to *evaluate* the background expectations and the current deliverances of any one sensory system by comparing them to the corresponding background expectations and current deliverances of a variety of independent maps. It can use any and all of its maps, and the fleeting activations within them, to keep any and all of its other maps honest.

There may thus be no hope, just as Kant wisely insisted, of getting a *map-independent* grip on the abstract content and structure of the objective and experience-transcendent world of things-as-they-are-in-themselves. There simply is no grip that would allow us to make a direct, fell-swoop evaluation of the extent and degree of the homomorphisms that may obtain between any given map and the objective domain it purports to portray. But with all this conceded, a map-*dependent* grip remains a real and a genuine grip, however grappled and groping might be its origins.

And a large family of independent but *overlapping* map-dependent grips, however feeble taken one by one, can collectively afford a very firm and trustworthy grip indeed, especially if they are allowed to foster ongoing adjustments in each other, over time, so as to maximize the mutual coherence of their individual portrayals of reality.

This, for better or for worse, is our own epistemological situation, at least according to the naturalistic, neural-network approach here being explored. That approach acknowledges our eternal dependence on the conceptual structures internal to whatever cognitive mechanisms we may bring to the business of world-apprehension. It is, in that respect, firmly Kantian. But at the same time, it denies the innateness and *implasticity* of (most) of those mechanisms. Kant notwithstanding, we are free to try to change and improve them, so as to enhance the integrity of the maps those cognitive mechanisms eventually acquire.

This naturalistic approach also asserts the possibility of *augmenting* our native cognitive mechanisms with entirely new cognitive maps, maps concerning the structure of hitherto inaccessible domains of reality, maps driven or indexed by entirely new instruments of detection and apprehension. Those new maps are called *scientific theories*, and they are given additional empirical life by such things as voltmeters, thermometers, Geiger counters, mass-spectrographs, and interferometers. In the hands of an adept, these new resources constitute an expansion of the original cognitive agent. Such an agent now commands an armory of *additional* presumptive 'grips' on objective reality, and will have, in consequence, an increased capacity for the business of comparing the deliverances of one empirically engaged map against the corresponding deliverances of many others.

Finally, and partly in consequence of the two points just made, the approach here proposed denies the Kantian Gulf said to be forever fixed between the 'Empirical World' on the one hand, and the 'World of Things-in-Themselves' on the other. We must indeed acknowledge the possibility—nay, the moral certainty—that our current portrayal of the world is radically incomplete in many dimensions, and flatly mistaken in many others. Though our modal enthusiasms may vary on this point, this concession, at least, is easily made by all parties to the debate. But the *world portrayed* by our acquired cognitive mechanisms remains always and ever the World of Things-in-Themselves, however pinched and inaccurate our current portrayal may happen to be. And the individual objects that we genuinely see, feel, and hear are always and ever the things-as-they-are-in-themselves, however idiosyncratic, partial, or twisted may be our current sensory and conceptual grip on them. Just as the Phenomenalists were wrong to interpolate an *inner* theater of subjective phenomena—distinct from physical objects—to contain the 'real' objects of our perception or cognitive apprehension, so was Kant wrong to interpolate an *outer* theater of spatiotemporal phenomena—distinct from Things-in-Themselves—to contain the 'real' objects of our perception or cognitive apprehension. Our problem is not that we perceive and interact with something *other* than Things-in-Themselves. On the contrary, we are up to our necks and ears in them. How could we possibly escape them? Our problem is only that we currently apprehend them with historically accidental resources that are inadequate to comprehend and track them in their full, and no doubt considerable, glory.

It is the job of scientific research, broadly conceived, to identify and chip away at those inadequacies, and to modify, amplify, or replace our current conceptual maps with better maps; and those, in time, with maps better still. Our opening lesson (detailed above) about the proper evaluation of existing maps is that there is no substitute for, and no alternative to, putting the map to work in a systematic attempt to navigate, in some fashion, the domain it purports to portray. This navigational effort must be aided by the simultaneous deployment of other activated maps that overlap the target map, in some potentially corroborative or critical ways.

Its cartographical metaphors notwithstanding, the reader will recognize in this inescapable lesson the familiar Pragmatist's criterion for the evaluation of our cognitive commitments. As a long-time student of the writings of C. S. Peirce, Wilfrid Sellars, and W. V. O. Quine, I confess that the lesson here (re)discovered arrives to antecedently willing ears. But there is an irony here, and a possible problem, that must be addressed. In confronting the issue of how to evaluate our cognitive commitments at any stage, Peirce himself was always inclined to steer our attention *away* from any direct or timeless comparison of those commitments with Reality itself, and *toward* the relative quality of the practical behavior or world navigation that those cognitive commitments actually produced. Even his official account of The Truth—as that theory on which all honest (and pragmatically guided) researchers will eventually converge—declines to appeal to any form of theory–world correspondence. Moreover, a fourth Pragmatist whom I admire, Richard Rorty, has argued explicitly that it is a false errand and a mug's game to try to explain one's accumulated wisdom about the world, or even one's perceptions of it, as being in any way an acquired *mirror* of Nature herself (see Rorty 1979). Forget its imagined 'reflective' virtues, he argues, and focus on what is its primary function in any case, namely, the ongoing generation of pragmatically successful interactions with the world.

Moreover, Peirce and Rorty have some up-to-date allies on this very point. The recent line of research that applies the framework and techniques of *dynamical systems theory* to the unfolding activities of the biological brain is as naturalistic and as pragmatically inclined as one could hope for (see Port and van Gelder 1995). But some of its leading proponents suggest that we should give up, because we no longer need, the idea that cognition ultimately consists in the exploitation and manipulation of 'representations' of any kind at all.[2] This suggestion, note well, is not a return to Classical Behaviorism, because we are not invited, not for a second, to regard the brain as an unapproachable black box. On the contrary, its internal dynamical mechanisms are the prime target of this theoretical approach. But the many regulatory functions that the brain undoubtedly performs for any creature—walking, foraging, burrowing, dodging, mating, and so forth—do not strictly *require* that its internal mechanisms or activities involve anything that answers to the category "representation." They do not require classical representations, such as *sentences* in some natural or computer language; and perhaps not geometrical representations either, such as the *activation vectors* and *sculpted activation-spaces* on display in the present book. All the brain requires, runs the suggestion, is a profile of internal dynamical behavior that produces appropriate motor behaviors in the perceptual environments it is most likely to encounter.

You will appreciate my potential embarrassment(s) here. I am suggesting that the brain's acquired maps are at least partially and approximately homomorphic with at least some of the objective feature spaces of things-in-themselves. But Kant will object that such homomorphisms are forever *impossible to evaluate*, or (more likely) that it makes no sense even to speak of them. Peirce and Rorty will both object that *one should not expect to find* any such reality-reflecting homomorphisms as the essence of the brain's acquired knowledge, since that essence properly lies elsewhere. And van Gelder will object that the brain *does not strictly need* any 'reality-reflecting maps' to perform its many regulatory functions, and he will point to successful regulatory systems (such as the whirling speed-governor on Watt's steam engine, perhaps) that *do not in fact deploy* any such maps.

2. T. van Gelder (1993), for example, provoked a flurry of discussion with his brief but penetrating newsletter article, "Pumping Intuitions with Watt's Engine."

My problem is that I admire all four of these philosophers. I count myself an advocate of almost everything they stand for, and I do not wish to opt out of the traditions that each one of them represents. Let me try, therefore, to present the thesis of the present book in such a fashion as to disarm the objections just surveyed, without rejecting wholesale each of the background traditions that give rise to them.

I address van Gelder's point first. He is correct in asserting that regulatory mechanisms need not, and often do not, use anything that is instructively seen as a map. But it remains true that many other regulatory mechanisms, some of them close at hand, plainly do. For example, there is a control box at the side of my house. It automatically turns on the electric motor that pumps water from our modest backyard pool through a filter. That device contains a clock-like face with twenty-four hours marked off on a scale around its circumference. That clock-driven face rotates once a day, past an unmoving pointer at the box's bottom center, which registers the current time by pointing at the slowly moving circular scale. Once more we have a background map with a 'current position' indicator in relative motion thereto. And this (circular) map of (daily) time has two adjustable clips on its rotating face, about 30° apart, clips that represent the two landmark times at which the control box turns the filter pump on, and then off again two hours later. As those clips go past the unmoving pointer, they each make an electrical contact. The first turns the pump-motor on, and the second turns it off. In this way is the pool water filtered for two hours every day.

Control devices like this are everywhere. An almost identical device turns on the ground-level electric lights in our backyard flower garden at eight o'clock every evening, and turns them off three hours later. Once again, a rotating temporal map with on/off landmarks does the trick. Yet a third device, this time digitally realized, does much the same thing for the in-ground lawn sprinkler systems so common in this near-desert region. Pools, gardens, and lawns aside, note well the ubiquitous wristwatch worn on every modern arm. Its background temporal map and moving pointer regulate the daily behaviors of almost every person on the planet.

The thermostat on your hallway wall is another variation on the same theme, only here the internal circular map-space relative to which a bimetal coil rotates back and forth is a map of the indoor *temperature* rather than of time. Once again, two landmark positions in that space turn the furnace on and then off again, as a map-specified minimum in room temperature is repeatedly detected and corrected for. A look at your car's speedometer, fuel gauge, and radio dial will reveal further variations on this same theme, although the maps deployed are maps of *velocity* space, *fuel-volume* space, and electromagnetic *frequency* space, respectively; and in this case, the driver himself is very much a part of the control-loop at issue. And while we are back on the topic of automobiles, recall again the entirely literal *road* map deployed by the onboard GPS guidance system in the autonomous vehicle discussed earlier, the one navigating its way around San Diego without any driver at all.

Evidently, deploying a background map of some practically relevant feature space, a map sporting some form of dynamical place-marker, is a common and highly effective technique for monitoring, modulating, and regulating many practically relevant behaviors. And the findings of modern neuroscience make it obvious, going in, that the biological brain makes at least occasional use of the same technique. The population of frequency-sensitive 'hair cells' lining the cochlea of your inner ear is plainly a metrically accurate map of acoustic frequency space, and their collective activity registers one's current power-spectrum position within that space. One's somatosensory cortex is a (metrically deformed) map of the body's external surface—literally, of one's skin. By contrast, one's motor cortex evidently embodies a (high-dimensional) map of the many complex limb-configurations one's body might assume (recall again Graziano, Taylor, and Moore 2002). The fly's olfactory bulb, as we saw earlier, is a high-dimensional map of odorant space. The human primary visual cortex is well known to be a (metrically deformed) topographical map of the retinal surface that projects to it, as are several other visual areas further up the processing ladder. That there are functional maps within the biological brain, at both the sensory and the motor ends of the system, is not a matter of any controversy. Nor is the *metrically deformed* character displayed by many of them, which hints immediately of map-to-map transformations as a natural computational technique.

It remains an entirely empirical question whether all or most of the brain's many other neuronal populations will turn out to harbor *further* maps—presumably of much higher dimensionality than the comparatively obvious (one- and two-dimensional) maps that dominate the set of peripheral examples just given. But it is a hypothesis well worth pursuing, given the obvious *map-transforming* prowess of the cadre of synaptic connections that allows any one neuronal population to activate another.

I am therefore strongly inclined to resist van Gelder's counsel that we simply forget about representations, and transformations upon them, entirely. The empirical evidence does indeed suggest that broadly 'syntactical' representations, and broadly 'inference-like' transformations upon them, play no role in animal cognition and only a minor and secondary role even in human cognition. That is a theme we share. But that very same evidence suggests that the forms of high-dimensional geometrical representation on display in this chapter play a prominent role in the cognition of every creature on the planet.

While I therefore reject this one aspect of van Gelder's counsel, I hasten to accept another. His preeminent message concerns the importance of the unfolding *temporal* dimension of cognition, a dimension deliberately sup-pressed, for purely expository reasons, in my discussions to this point. That dereliction will be repaired starting in the next section, and at length in the following chapter. What will there emerge is that, when we add, to our theoretical picture, the vitally important *recurrent axonal projections* that characterize any biological brain, the resulting picture is a highly fertile *instance* of the dynamical systems approach urged by Port, van Gelder, and so many others. It is in no way opposed to that approach. Accordingly, vectorial representations and high-dimensional maps should not be shunted aside along with the parochial syntactic representations of classi-cal AI. On the contrary, they appear to be an integral part of the avowedly dynamical story that needs to be told.

The theme of the preceding paragraphs is also the beginning of my response to Peirce and Rorty. Insofar as they are both concerned to reject any classical Correspondence Theory of Truth—where the parties to the relevant correspondence are the set-theoretical structures embodied in the world, on the one hand, and the syntactic structures displayed in certain preferred sets of interpreted sentences, on the other—then the present book is in substantial sympathy with both thinkers. For that is the *wrong place to look* for the global correspondences that give any brain its primary grip on the world. Most cognitive creatures—that is, every cognitive crea-ture on the planet except us—have no command at all of the syntactic structures required by a Tarski-style account of Truth. And even for us adult humans, our linguaformal representations play only a secondary role in our cognition in general. This should be no surprise: humans, too, are animals. And our brains are only slightly different, both qualitatively and quantitatively, from the brains of the other primates, and from the brains of other mammals, for that matter. Evidently, the classical Correspondence Theory of Truth is an utterly hopeless account of the cognitive achieve-ments of terrestrial-creatures-in-general.

But the genuine absence, or the poverty, of any grand explanatory cor-respondences, between the two *classical* domains just specified, need not mean that there are *no* grand explanatory correspondences between the quite different domains specified by an adequate account of how the *biological brain* represents the world. If we take, as our (potentially) corresponding domains, the family of objective similarity and difference relations that unite and divide the prominent elements of some objective feature space, on the one hand, and the high-dimensional feature-space map embodied in some suitably instructed neuronal population, on the other, then we confront two immediate rewards. First, we have an account of how cognitive creatures represent the world that is not absurdly restricted to creatures that command a human-like language. And second, we have an intelligible account of what sort of correspondence underlies representational success: it consists in precisely those kinds of homomorphisms that make any structure an *accurate map*.

There is a third major payoff here, one of especial interest to anyone with Pragmatic inclinations. The brain's possession of behavior-guiding, high-dimensional, feature-space maps allows us to construct illuminating explanations of our occasional behavioral *mis*adventures. A map, after all, may be inaccurate in some one or more particulars. Here and there, it may fail of a perfect homomorphism with its target domain. And when the map's unsuspecting owner wanders into one of those misrepresented portions of the target domain, the brain will be led to inappropriate expectations of what the world holds in store for it, and to false takes on the local behavioral opportunities that the world affords. Disappointment is then inevitable, as when our autonomous GPS-guided automobile is abruptly brought up short against a steel barrier as it attempts to navigate a complex freeway interchange that was redesigned entirely and rebuilt three years ago. Its computer-stored map, not yet appropriately updated, has led the car astray.

Chronic misrepresentations aside, fleeting misrepresentations are also possible, as when the GPS locater-system wrongly indicates that the car is moving west on Lomas Santa Fe Drive at the Pacific Coast Highway 101, when it is actually moving west all right, but straight into the pounding surf, on the beach at Fletcher's Cove, 200 yards away. Here the background map need have no representational defect. Rather, it is the sensory system's ephemeral activation or registration of a false location within that background, 200 yards to the east of where it should be, that leads to the trouble.

We can thus hope to give highly specific explanations, of the diverse disappointments and multiform misadventures encountered by any creature, in terms of either its enduring mis*con*ceptions of the world, on the one hand, or its ephemeral mis*per*ceptions of the world, on the other. We can find specific fault with its representations at either one of these two distinct levels. And having located such faults, whether in ourselves or in others, we can have some hope of repairing them or avoiding them in the future.

By the same token, and no less than failure, behavioral *success* is also brought within explanatory reach. The navigational success of our driver-less automobile is a startling thing, until we are advised of both its onboard map of San Diego's streets, and its continuously updated GPS-driven self-location within that map. Its navigational feat, while still impressive, is then nicely explained. Similarly, the navigational success of any animal—a spider, a mouse, an owl—would be a comparably startling thing, save for our appreciation of its internal maps of the several feature-spaces charac-teristic of its environmental niche, and its several sensory systems for registering a current feature-space position within those maps. The diverse successes of those creatures, while still impressive, are thus nicely and systematically explained.

Thus we have a case for insisting on a conception of, and at least a comparative measure for, cognitive or representational *virtue* that is inde-pendent of any measure of pragmatic or navigational success. For we want to be able to *explain* the profile of any creature's pragmatic successes and failures *in terms of* the profile of virtues and vices of its various background representational systems. If, on the other hand, we choose to *define* "truth" or "representational virtue" directly in terms of pragmatic success—as in, "the true is what works"—we deny ourselves all access to an evidently rich domain of potential explanations.

This leads us, plausibly and prosaically enough, back to a recognizable form of Scientific Realism. Specifically, an ever-plausible (but ever-defeasible) explanation for the predictive, explanatory, and manipulative superiority to date, of some conceptual framework *A* over some other con-ceptual framework *B*, is that *A* embodies the more accurate representation of the objective domain in which that comparative pragmatic evaluation took place. To be sure, 'conceptual frameworks' here are 'high-dimensional feature-space maps' rather than scientific theories as classically conceived. But that is a gap that will be closed as these chapters unfold.

Finally, and most generally, there is the Kantian-style objection that such partial or approximate homomorphisms as might obtain between brain maps and objective feature spaces are forever beyond human evalu-ation, and are therefore beyond the bounds of meaningful discussion. The short answer is that both the existence and the extent of such homomor-phisms are very much within the range of human evaluation. For example,

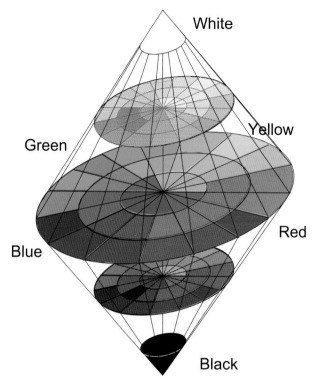

White

Yellow

Green

Blue

Red

Black

Plate 1
A map of the space of possible colors.

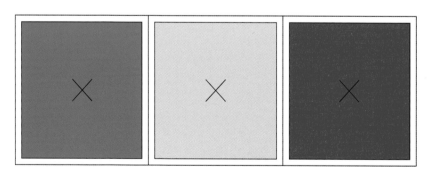

Plate 2
Colored afterimages due to neuronal fatigue.

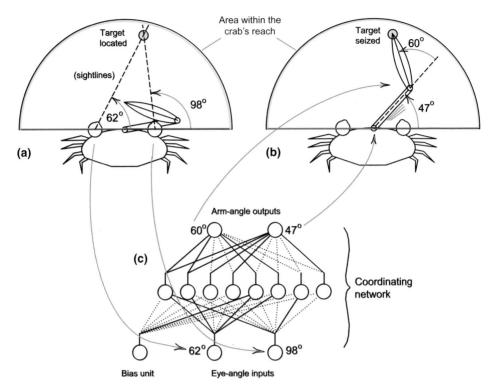

Plate 3
Sensorimotor coordination achieved by a map-transforming neuronal network.

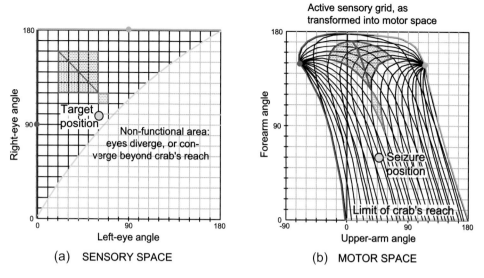

Plate 4

The metrical deformation of active sensory space into active motor space.

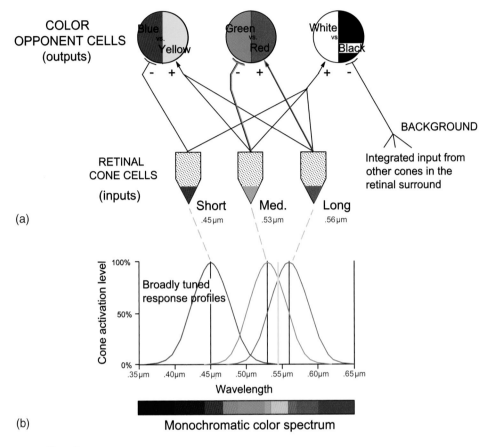

COLOR
OPPONENT CELLS
(outputs)

Blue vs. Yellow

Green vs. Red

White vs. Black

BACKGROUND

RETINAL
CONE CELLS
(inputs)

Integrated input from
other cones in the
retinal surround

(a)

Short
.45 μm

Med.
.53 μm

Long
.56 μm

(b)

Cone activation level

100%

50%

0%

Broadly tuned
response profiles

.35 μm   .40 μm   .45 μm   .50 μm   .55 μm   .60 μm   .65 μm

Wavelength

Monochromatic color spectrum

Plate 5
The human color-processing network (after Jameson and Hurvich 1972).

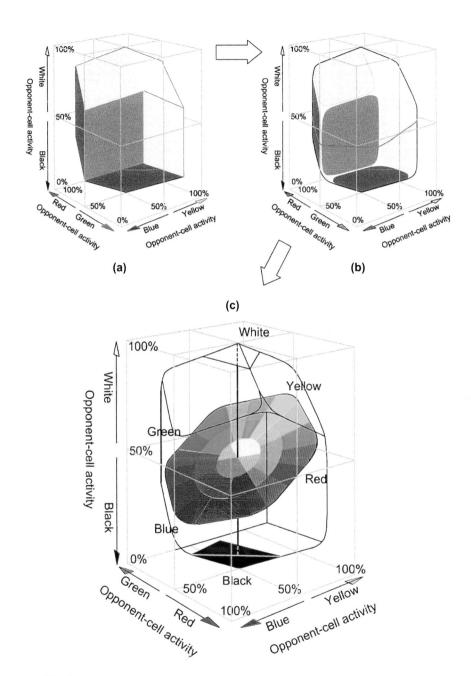

Plate 6
The solid neuronal activation space of the Hurvich-Jameson network, and its internal map of the colors.

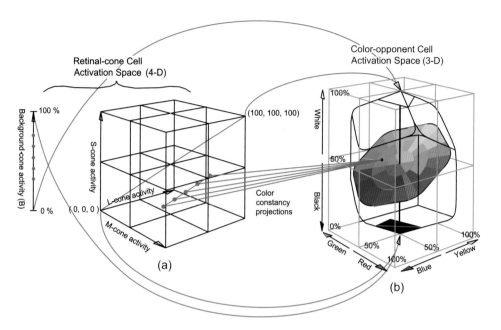

Plate 7

Color-constancy in the Hurvich-Jameson network.

Plate 8
The slippery slope to Idealism (my thanks to Marion Churchland).

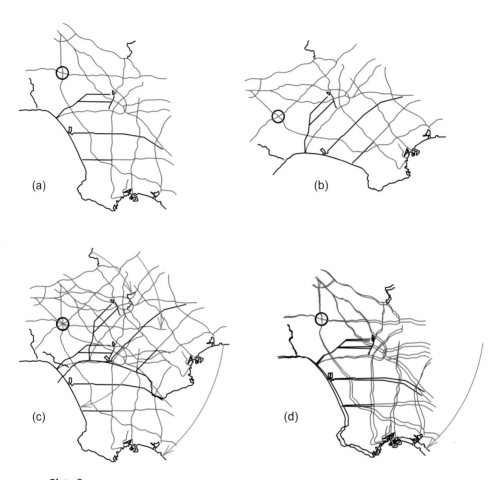

(a)

(b)

(c)

(d)

Plate 9
Two distinct maps superimposed and rotated to find a systematic homomorphism.

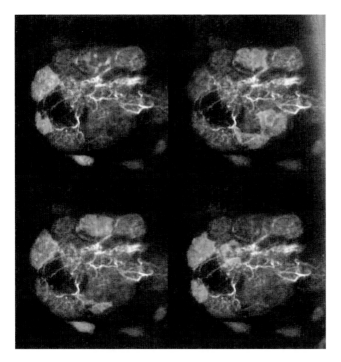

Plate 10
Distinct activation patterns in the fly's olfactory glomeruli. (Courtesy of J. W. Wang, Center for Neurobiology and Behavior, Columbia University, New York, USA.)

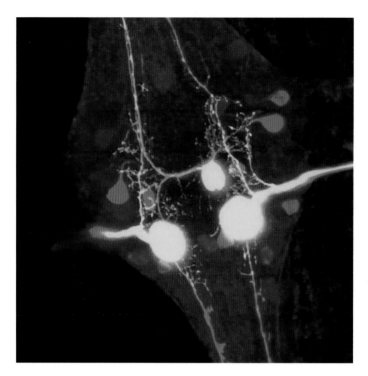

Plate 11
Motor neuron activity in the leech. (Courtesy of W. Kristin, Department of Neuroscience, UCSD, La Jolla, CA, USA.)

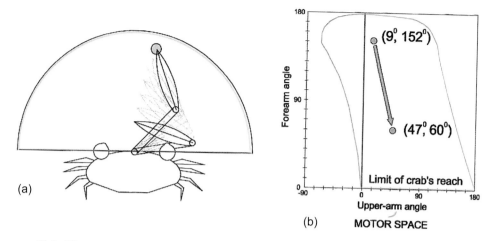

Plate 12

(a) A continuous motion of the crab's arm, as represented in (b) the crab's motor space.

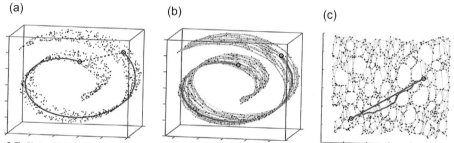

Fig. 3. The "Swiss roll" data set, illustrating how Isomap exploits geodesic paths for nonlinear dimensionality reduction. **(A)** For two arbitrary points (circled) on a nonlinear manifold, their Euclidean distance in the high-dimensional input space (length of dashed line) may not accurately reflect their intrinsic similarity, as measured by geodesic distance along the low-dimensional manifold (length of solid curve). **(B)** The neighborhood graph G constructed in step one of Isomap (with $K = 7$ and $N = $ 1000 data points) allows an approximation (red segments) to the true geodesic path to be computed efficiently in step two, as the shortest path in G. **(C)** The two-dimensional embedding recovered by Isomap in step three, which best preserves the shortest path distances in the neighborhood graph (overlaid). Straight lines in the embedding (blue) now represent simpler and cleaner approximations to the true geodesic paths than do the corresponding graph paths (red).

Plate 13

A further technique for the dimensionality reduction of high-dimensional sensory inputs.
(Thanks to Tenenbaum, de Silva, and Langford [2000]. Reprinted with permission.)

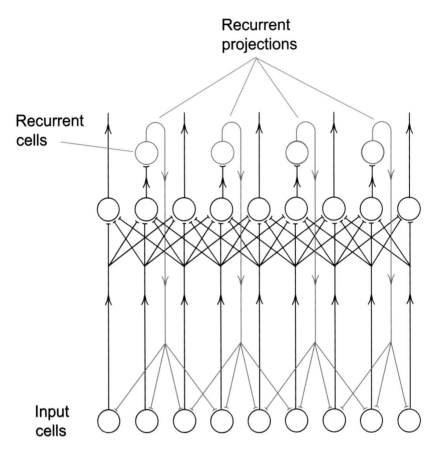

Recurrent
projections

Recurrent
cells

Input
cells

Plate 14
A simple recurrent network.

external structure of the objective similarity and dissimilarity relations that configure the range of human faces on the other, is easily seen and is traceable in some detail, at least where the representation of gender is concerned.[3]

To switch examples, each of the three rungs in the crab-coordinating network of chapter 2, section 3, embodies a distinct map—of eye-angle space at the input level, of sensorimotor-coordination space at the middle level, and of arm-position space at the output level. The homomorphism between the first rung's two-dimensional sensory map and the objective range of possible joint eye-rotations is simple and transparent, as it also is between the third rung's two-dimensional motor map and the objective range of possible joint arm-angles. (See again fig. 2.6a,b. Note also that *these* two maps were not learned, but were deliberately 'hardwired' into the network at the outset.)

Less obviously, the eight-dimensional map at the middle rung, the one slowly sculpted by the back-propagation algorithm, also boasts a homomorphism—in this case, with an abstract space of specific <eye-angle pair, arm-angle pair> *transformations*. That map does not represent all possible transformations. As with so many other activation spaces, the learned map embraces only a smallish part of the total eight-dimensional space that is initially available. In this case, the learned map embraces only a special subset of all possible input-to-output transformations: namely, the ones that place the tip of the crab's pincer exactly at the intersection of the crab's two sightlines, wherever those two sightlines happen currently to be. The middle-rung coding points that effect those highly special transformations are all located on a curved two-dimensional surface within that eight-dimensional space. And the points on that surface are topographically organized to mirror the topographic organizations of the points in both eye-angle space and arm-angle space (that is to say, *neighborhood*

---

3. It is rather less obvious what learned internal dimensions reflect, in the same way, the several objective facial differentia that divide the eleven human *individuals* in the training set, and allow the reidentification of the same individual across distinct photos. They are likely to be more subtle than the comparatively gross differences that characterize gender. And those same internal dimensions are likely to be peculiar to the randomly drawn group of eleven people whose photos made up the original training set. That is to say, training a second network on an entirely different group of faces will likely produce a slightly different map of interfacial similarities and differences, a map that betrays itself by producing a slightly different profile of discriminatory behavior across the two networks. Precisely this result has been documented in humans by O'Toole, Peterson, and Deffenbacher (1996).

we can plainly evaluate the general accuracy, and the specific f.
an outdated AAA highway map, and the case of a higher-dimensio.
map differs in no fundamental respect from that mundane case.

To illustrate this evaluative accessibility with an unproblematic ex
return your attention to the face-recognition network discussed earl
succeed in making reliable judgments as to the gender and name
eleven faces in its training set, the network had to become sensitive
least some of the objective differences that typically distinguish a fer
face from a male face. (Fixing on hair length won't do it—not here, no
the population at large—for two of the males in the training set had lo
shoulder-length hair, and two of the females had closely cropped hai
Those several objective and enduring differences turn out to include,
noted earlier, (1) the vertical distance between the pupil and the eyebrow
it is usually larger in females; (2) the vertical distance between the bottom
of the nose and the upper lip: it is usually smaller in females; and (3) the
vertical distance between the mouthline and the bottom of the chin: it is
usually smaller in females. Speaking roughly, and summarizing from the
male perspective, a typical male face is a beetle-browed, lantern-jawed
deformation of an otherwise perfectly good human face.

In the course of training, the network did indeed become sensitive to
precisely these sorts of variations across faces, a fact reflected in the substantial distance separating its activation-space positions, at the second
rung, for coding prototypical male and prototypical female faces, respectively (see again fig. 1.4). They are on opposite sides of the relevant activation space, roughly equidistant from its geometrical center. By contrast, a
gender-neutral or gender-ambiguous face will find itself coded at some
point roughly equidistant from both of those prototype points—for
example, at the midpoint of a straight (hyper)line between them, although
there are many other possibilities as well, all located on the (hyper)plane
that intersects that midpoint and lies normal to that line.

Further, human faces that are either hyperbolically male or hyperbolically female—as measured by the three parameters cited above—will find
themselves coded somewhere in the extremal regions *beyond* the male and
female prototype points. Those outliers will be coded at some point or
other, some maximal distance away from the central (hyper)plane of
gender neutrality that divides the male-face region from the female-face
region. See again the range of objective faces, and their corresponding
coding positions within the second-rung face map, in figure 1.4.

In this case, the homomorphism between the internal structure of the
activation space's acquired similarity metric on the one hand, and the

relations are preserved through both transformations). Points outside of that preferred surface are never activated, save during some neuronal malfunction, and they effect dysfunctional motor outputs if and when they are activated. This preferred surface is a map of all and only the coordinatively *successful* transformations. All others are simply ignored.

Finally, return your attention to the familiar three-dimensional map of objective color space (fig. 1.3, plate 1). This case is interesting for many reasons, not least because a number of contemporary thinkers have argued, perhaps cogently, that objective colors are a grand illusion. The argument proceeds by claiming that, as a matter of scientific fact, nothing in the objective world *answers* to the peculiar family of similarity and difference relations so robustly portrayed in our native color space. That internal map is judged, by these thinkers, to be an inaccurate map, a false map, an outright misrepresentation of the real structure of the objective, light-reflecting, physical reality (see, e.g., Hardin 1993). Note that this complaint is distinct from Locke's, who lamented the lack of *first*-order resemblances between our inner states and outer properties. The present complaint laments the alleged lack of any appropriate homomorphism or *second*-order resemblance between our internal map and the objective, external feature-space we confront (namely, the range of electromagnetic *reflectance profiles* displayed by physical objects). Whereas Locke's complaint was confused and inapt, as we saw above, this newer complaint may have a genuine point.

Whether or not it has, I have explored in an earlier paper (Churchland 2007b), and I will spare you the details here. But either way, the complaint illustrates both the possibility of false maps, and the possibility of unmasking their specific failures by comparing their internal structures with the structure of *other* maps—maps that overlap the same domain, maps supplied, perhaps, by new scientific theories, maps that get activated or indexed by novel instruments of measurement and detection. In the case of color, the new theory concerns the reflection and absorption of electromagnetic radiation, and the novel instruments are spectrometers and photometers.

Kant is surely correct in insisting that we have no hope of ever stepping away from our accumulated cognitive machinery, so as to gain a direct 'transcendental peek' at how things-in-themselves 'really are'; for there is no such 'transcendental perspective' for us, or anyone else, to aspire to. But we can do several worthwhile things short of 'achieving' that false and empty goal. First, and *pro tem*, we can take the integrity of our current maps for granted, and then seek to understand and evaluate the cognitive strategies, tactics, and maps of other creatures like ourselves, both bio-logical and artificial. Second, and lifting the *pro tem* assumption just mentioned, we can play our several native cognitive maps off against one another in their areas of mutual overlap, progressively modifying each map in response to the need to maintain consistency and consilience across the entire set. And third, we can create entirely new cognitive maps in the form of new scientific theories, maps that get fleetingly indexed by measuring instruments above and beyond the humble biological senses bequeathed to us by the human genome. These theories can also be evaluated for representational accuracy, both by the unfolding quality of the pragmatic world-navigation that they make possible, and by the straightforward comparison of several distinct but overlapping maps with one another, in pursuit of mutual criticism, modification, and potential unification.

Accordingly, the world of things-in-themselves is not inaccessible to us at all. Indeed, it is the constant target of all of our cognitive activities. Our current representations of it are doubtless far from perfect, and may remain imperfect, to some degree, in perpetuity. But we are always in a position to make them better. And presumptively we have made them better. Kant himself would be astonished, and delighted, with what we have learned about the objective universe since he last deployed his own naïve cognitive maps upon it, secure in his equally naïve assumption that they guaranteed the truth of Euclidean geometry, the distinction between time and space, and the causal connectedness of all events.

In particular, we have learned that space is not Euclidean, at least in the neighborhood of large masses. And time is not distinct from space, a fact that becomes empirically obvious at high relative velocities. And the law of universal causal community fails dramatically at the level of quantum phenomena. These celebrated results, and others, both behind us and presumably ahead of us in history, speak to the poverty of Kant's curiously ahistorical conception of human cognitive activity, and to the profound mutability of human cognition. Granted, it may take substantial and pro-tracted *effort* to rise above Kant's classical convictions, and learn to think and to perceive within a new framework. But it is clearly possible: physicists, astronomers, and mathematicians use non-Kantian frameworks on a daily basis. Such conceptual retooling is also mandatory for any serious investigator, given the scientific evidence accumulated during the twentieth century. Representational frameworks, I conclude, are both variable across time, and evaluatable against more than one external standard. The conjunction of these two facts means that intellectual progress is at least robustly possible, if still not guaranteed.

2   The Neuronal Representation of Structures Unfolding in Time

Faces are not usually so obliging as to sit, upright and motionless, a constant distance in front of one. Instead, they rotate variously back and forth, as their attention is drawn to the left or right (fig. 3.1). They tilt forward and backward, as they look down at the desk or up at the light. They loom large as they walk toward one, and shrink again as they back away. They smile and frown and squint and move their lips in conversation. They behave nothing like the frozen snapshots that exhaust the experience, and also the comprehension, of Cottrell's classic face-recognition network. And yet the human brain sees unity in all of the flux just described, and more. It sees a unitary, head-shaped solid object, with an occasionally plastic front half, moving smoothly through space in a limited variety of characteristic ways, while displaying a coherent sequence of socially significant behaviors across its frontal features. What representational resources allow

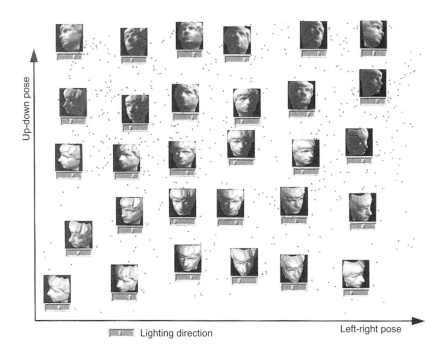

Figure 3.1
The variety of distinct sensory representations of the same face under different conditions of viewing. (Thanks to Tenenbaum, de Silva, and Langford [2000]. Reprinted with permission.)

the brain to get a grip on these complexities, and in particular, on their characteristic temporal and causal structures?

The behavior of moving faces barely hints at the magnitude of the question here confronted, for we must ask how the brain gets a grip on causal processes generally, as they are passively perceived in the physical world. Rolling balls, arcing arrows, swinging pendulums, flying birds, running quadrupeds, swimming fish, flowing water, flaring fires, and reciprocating saws all pose a similar challenge to our representational resources, and all of them are handled with evident grace and ease. Such prototypical processes are perceptually recognized very swiftly. How do we do it?

One possibility, classically embraced, is that their recognition depends on the brain's prior or independent recognition of the specific spatial character of the object or substance undergoing the process at issue, a recognition then augmented by a series of subsequent recognitions of spatial changes in that object over time. This initially plausible suggestion is almost certainly mistaken. For one thing, it privileges the perception of purely spatial structure over the perception of an undivided spatiotemporal structure, a priority that is dubious, given the ubiquity of causal processes in the perceptual life of any creature, even a young one, and given the often life-and-death importance of discriminating, as swiftly as possible, one such process from another. Indeed, it may be that enduring spatial particulars are a subsequent *abstraction* from the brain's previously achieved grip on certain kinds of unfolding causal processes, rather than the other way around.

For another thing, we already know that objects that are too poorly or too ambiguously perceived to be identified in a motionless snapshot are often swiftly recognized from their characteristic gait or behavioral profile, even if only dimly apprehended. What might have been a small, gray submarine twig three feet beneath the creek's surface is suddenly seen, thanks to its signature undulation, as a trout keeping station against the gentle current. What might have been a sunlit dust-mote, drifting out of focus across your field of vision, is recognized instead as a tiny gnat by its discontinuous zig-zag path. What might have been a bean bag, hurled over the fence from your neighbor's back yard, can be seen to be an arcing sparrow from the barely visible 8 Hz flicker of its tiny beating wings.

Our facility in recovering "structure from motion" is strikingly illustrated in a famous psychophysical experiment (Johansson 1973). Two complex interacting objects were each outfitted with roughly two dozen small lights attached to various critical parts thereof, and the room lights were then completely extinguished, leaving only the attached lights as effective visual stimuli for any observer. A single snapshot of the two objects, in the midst of their mobile interaction, presented a meaningless and undecipherable scatter of luminous dots to any naïve viewer. But if a movie film or video clip of those two invisible objects were presented, instead of the freeze-frame snapshot, most viewers could recognize, within less than a second and despite the spatial poverty of the visual stimuli described, that they were watching two humans ballroom-dancing in the pitch dark. The collective movement of those tiny lights—placed at the ankle, the knee, the hip, and so on—made unique and almost instant sense on the perceptual interpretation indicated, including the fleeting occlusion of some of those lights as the dancers whirled 'round each other. Everyone recognized this unfolding causal process within a second or two.

Perception aside, we also have to account for the brain's capacity to *generate* the same sorts of complex and coherent motor behaviors that it so effortlessly recognizes in other agents. This generational capacity is often displayed within minutes of birth, before spatial pattern-recognition has had much of a chance to set up shop. Think of the turtle, or the chick, ambulatory right out of the egg. Even the newborn gazelle—a higher mammal—is ambulatory within twenty minutes. Plainly, temporal structure is an integral part of neuronal representation, even at its earliest and most formative stages.

To meet this fundamental family of representational demands, the brain of any creature needs to learn not so much to activate prototypical *points* within its sundry activation spaces, as in all of the examples we have discussed so far, but rather to activate prototypical *trajectories* or *paths through* those activation spaces. To grasp this point quickly, consider again the trivial but deliberately transparent example of the crab reaching out its jointed arm, from an initial position folded against its body, to an extended position that makes contact with a visually located target (fig. 3.2a). That specific kinematic sequence, one of endless possible motions for that arm, is uniquely and timelessly represented, within the crab's motor output space, by the red *trajectory* indicated in figure 3.2b (plate 12). The time course of that motion will be represented by the actual time taken for an activation point in that space to traverse that trajectory from its initial point to its end point. A longish time represents a leisurely motion; a short time represents a swift motion.

But let us return to a more realistic and rather more demanding case— the case of the moving head as displayed in figure 3.1. Here we have a perceptual example: a large but finite population of perceived human-head positions within a continuum of possible human-head positions. Each position is initially represented by a single point in a high-dimensional 'retinal' grayscale space of (64 × 64 pixels =) 4,096 dimensions: that is, by

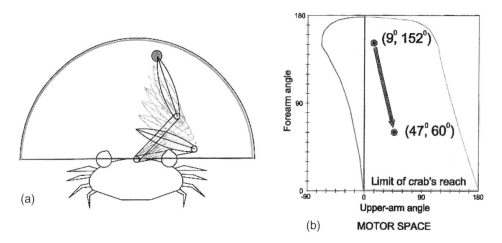

Figure 3.2
(a) A continuous *motion* of the crab's arm, as represented in (b) the crab's motor space. See plate 12.

one of the many pictures displayed in figure 3.1. (This 'pixel space' is a high-dimensional space, to be sure, but note that it is still small compared to the activation space of the human retina, which boasts almost 100 million pixels or light-sensitive rods and cones.) We now need to ask, is there any more frugal way to represent the underlying reality and its characteristic variations as portrayed in these many pictures? Indeed there is. You and I already appreciate that, in the many and various pictures of figure 3.1, we are looking at a unitary solid object, shaped like a human head, which varies in only three ways: (1) in its left–right rotation about a vertical axis through the top of its head, (2) in its tilt-up/tilt-down rotation about a horizontal axis through its ears, and (3) in the left–right variation in the source of its presumed illumination.

Note well that this appreciation, or recognition, of the underlying reality constitutes another case of *dimensionality reduction* or *information compression* of the sort already seen in Usui's color network and Cottrell's face network. Here we have 4,096 dimensions of variation (pixel brightness) reduced to only three dimensions of variation, those portrayed in figure 3.1.[4] Each illuminated head position is represented by a single point in that more frugal three-dimensional space. The various 4,096-pixel

4. Of course, that figure is a two-dimensional print, but note the cursor on the horizontal bar under each image: its position portrays the relevant third dimension, viz., lighting angle.

pictures plastered across figure 3.1 provide sample illustrations of which unique high-dimensional vector gets transformed into which low-dimensional daughter vector within the low-dimensional space there portrayed.

But how can any creature, or artificial network, ever acquire or *learn* this highly specific dimension-reducing transformation? In any number of different ways, it appears. Most obviously, we could simply construct a simple two-rung network, with 4,096 units at the input population projecting to only three units at the second rung, and then train that network, using the supervisory back-propagation algorithm discussed earlier, on a large population of appropriate input–output pairs deliberately drawn from the transforming function at issue. That transforming function is strongly nonlinear, as we shall see, but if the neuronal units of our network have sigmoid output-response profiles (see again fig. 2.4), we can approximate the desired function fairly closely. (An intervening middle rung might help in this regard. See again the first three stages of the five-rung Usui network.)

And yet, this approach is unsatisfactory in at least two respects. First, it requires that whoever is training the network (1) *already knows* the dimensionality of the underlying reality (three), so as to restrict the network's target rung to exactly three neurons, and (2) *already knows* the transforming function at issue, so as to provide the needed training instances of exactly that function, and (3) *repeatedly measures* the student network's successive approximations to them. The approach described is thus an instance of *supervised, error-reducing* learning.

Realistically, however, creatures in the wild have to learn on their own, with no antecedent or independent access to the 'right answers.' We plainly need to explore the prospects for some sort of *unsupervised* learning. The approach at issue (back-propagation) is also defective in a second respect: it requires that the network's synaptic connections be progressively modified by an elaborate error-reducing procedure that is biologically unrealistic in any case. It is too pointillistic; it is far too slow, especially for large networks; and it requires computations not obviously available to any nervous system.

One can wiggle on the first hook—as Usui and Cottrell both do, with real success—by turning to an *auto-associative* network with a deliberately frugal middle rung. The network is thus forced to *search* for a successful dimension-reducing transformation (and a dimension-expanding inverse) without anyone knowing, or needing to know, what those two transformations are beforehand. This is a neat and suggestive trick. And it does strictly dispel the need for a relevantly *prescient* supervisor. (Any idiot, and any program, can recognize instances of, and departures from, the identity transformation, which is all one needs to back-propagate the errors discovered in an *auto-associative* network.) But the second hook remains, and looms large: the back-propagation learning algorithm is a nice enough technology, but it is not Nature's technology.

Despair would be premature, however, for there are at least two further demonstrably effective procedures that will solve the sort of dimension-reduction problems here at issue, without illegal prescience, given sufficient examples from which to learn. Two recent reports (Tenenbaum, de Silva, and Langford 2000; Roweis and Saul 2000; see also Belkin and Niyogi 2003) outline two closely related but distinct procedures that will find, without any prescient supervision, a low-dimensional subspace within the given high-dimensional data space, a subspace that represents the objective dimensions of variation within the underlying reality, the reality that actually generated the peculiar set of data points encountered in the initial high-dimensional data space. And both of these dimension-reducing procedures will succeed even if the transformation from the higher to the lower space is nonlinear, a virtue that allows them to transcend the linear limitations of the techniques, already familiar to mathematicians, of principal components analysis (PCA) and multidimensional scaling (MDS).

Figure 3.3a (see plate 13) portrays, in a low-dimensional cartoon form, the set of given data points (e.g., the many 4096-D vectors embodied in the many pictures of the rotating plaster head) as they are plotted in that initial 4096-D space. Figure 3.3b explicitly portrays the discovered low-dimensional nonlinear manifold, embedded therein, which represents the objective dimensions of variation suffered by the rotating plaster head. And figure 3.3c portrays that subspace once more, now freed from its original embedding and unrolled into a transparently Euclidean manifold. Therein, the distinct coordinate-pairs can represent objective head-positions by their objective rotation angles around a single vertical axis and a single horizontal axis, instead of by their original 4,096-pixel grayscale presentations in a high-dimensional 'appearance space.' (Just to remind: the cartoon of fig. 3.3 portrays a 3-D to 2-D reduction. The actual reduction at issue here [fig. 3.1] was a 4,096-D to 3-D reduction.)

A second illustration of what these dimension-reducing techniques can achieve appears in figure 3.4 where the given high-dimensional data points are again pictures with 60×60 pixels. The algorithms at issue successfully recover the variations in the underlying low-dimensional reality, namely, a hand and wrist that vary in degree of wrist rotation, degree of finger-extension, and lighting angle. A welcome feature of both cases of

(a)   (b)   (c)

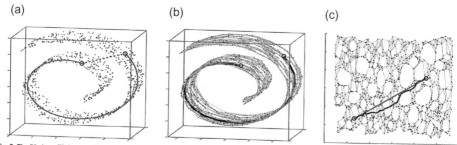

Fig. 3. The "Swiss roll" data set, illustrating how Isomap exploits geodesic paths for nonlinear dimensionality reduction. (A) For two arbitrary points (circled) on a nonlinear manifold, their Euclidean distance in the high-dimensional input space (length of dashed line) may not accurately reflect their intrinsic similarity, as measured by geodesic distance along the low-dimensional manifold (length of solid curve). (B) The neighborhood graph G constructed in step one of Isomap (with $K = 7$ and $N =$ 1000 data points) allows an approximation (red segments) to the true geodesic path to be computed efficiently in step two, as the shortest path in G. (C) The two-dimensional embedding recovered by Isomap in step three, which best preserves the shortest path distances in the neighborhood graph (overlaid). Straight lines in the embedding (blue) now represent simpler and cleaner approximations to the true geodesic paths than do the corresponding graph paths (red).

Figure 3.3

A further technique for the dimensionality reduction of high-dimensional sensory inputs. (Thanks to Tenenbaum, de Silva, and Langford [2000]. Reprinted with permission.) See plate 13.

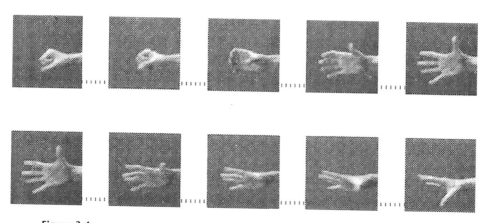

Figure 3.4

Two complex motion-sequences: the human hand. (Thanks to Tenenbaum, de Silva, and Langford [2000]. Reprinted with permission.)

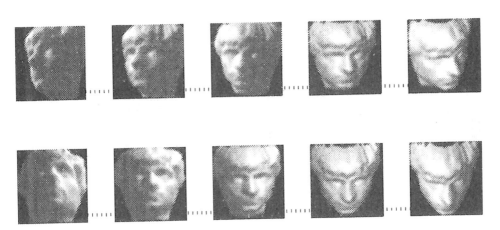

Figure 3.5
Two complex motion-sequences: the human head. (Thanks to Tenenbaum, de Silva, and Langford [2000]. Reprinted with permission.)

low-dimensional representation is that a continuous path confined to the low-dimensional subspace represents a continuous *motion* possible for the object therein represented. Different paths represent different possible motions for that object, with different possible starting and finishing positions. This is what we see in the two hand-sequences of figure 3.4 and in the two head-sequences of figure 3.5. Indeed, *any* possible motion in objective space, for either the head or the hand, is represented by some continuous path or other within the relevant low-dimensional subspace.

(It is the existence of such representationally insightful low-dimensional subspaces that I wish to emphasize here, as opposed to whatever algorithmic or biological procedures might be used to pull them out of their murky high-dimensional embedding spaces. Those procedures are the topic of the next section.)

Evidently, such transformation-mediated lower-dimensional spaces provide a most effective medium for representing the essential features of any *structured process unfolding in time*. A *closed* path in such a space, to cite one important class of examples, can represent a rhythmic or periodic motion such as walking, flapping, swimming, chewing, or scratching, where the relevant process leads repeatedly back to its own starting point. Moreover, and just as important, such high-dimensional to low-dimensional transformations reflect, in a revealing way, the distinction between the complex and perspective-dependent *appearances* presented to a cognitive system on the one hand, and its subsequent grip on a simple and perspective-independent *reality* on the other, where both the appearance and the reality involve complex processes unfolding in time.

The principal suggestion to this point is that prototypical bodily motions and other causal *processes* unfolding in time can, at least in principle, be efficiently and comprehensively represented by prototypical *trajectories* or *pathways* through the neuronal activation space of a population of neurons, specifically, by pathways within a learned, comparatively low-dimensional submanifold that represents the basic variations within the enduring underlying features discovered for the motion or process at issue. Similar motions get represented by proximate pathways within that special submanifold. Dissimilar motions get represented by quite different pathways. Motions that share a common physical configuration at some point get represented by pathways that intersect at some point. Repetitive motions get represented by pathways that form a closed loop—a *limit cycle* or stable oscillation. Collectively, the many trajectories within that special submanifold represent *all possible* motions or causal processes of the type therein comprehended.

So far, however, the vector-algebraic algorithms that assure the abstract existence of such representational submanifolds, and pathways within them, do nothing to tell us how to realize such a manifold in a neuronal system so as to permit the appropriately sequential *activation* of the points along any such prototypical pathway. A capacity for initiating and pursuing such prototypical sequences of activation points will be essential if the neuronal system is ever to use such preferred trajectories to generate real motor behavior, for example. And such a capacity will be equally essential for activating some such preferred trajectory so as to achieve perceptual recognition of some prototypical motion or causal process, as a brain-generated anticipatory *completion* prompted by the perception of the motion's characteristic first few stages. Specifically, we need a style of neuronal network that learns to exhibit a family of prototypical *sequences* of activation points in response to a given stimulus, instead of exhibiting a family of mere prototypical *points*, as in all of the network examples discussed so far in this chapter. We need a neuronal system that will 'play a brief movie' as opposed to 'display an occasional snapshot.'

As it happens, the required capacity arises naturally in a straightforward articulation of the purely feed-forward style of network familiar from the many examples already discussed in this chapter. The critical addition is that a significant percentage of the neurons at some rung of the relevant processing ladder be induced to project their axons *downward* (see fig. 3.6, plate 14)—to make systematic synaptic connections with the neurons at

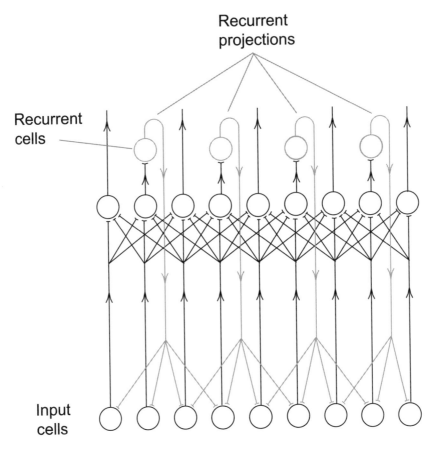

Figure 3.6
A simple recurrent network. See plate 14.

some rung *lower* in the processing hierarchy—instead of projecting only upward, always upward. These 'descending' or 'reentrant' or 'recurrent' axonal projections—as they are variously called—give the higher rung from which they originate a significant degree of *control* over the activities of the neurons in the lower rung that receives those special projections.

Their presence also means that the ways in which that lower rung responds to any sensory input, from still lower in the ladder, is no longer determined solely by the character of that sensory input and by the configuration of synaptic weights through which that input must pass. For the neuronal response at that now doubly connected rung will be partly a function of the prior activational state of the higher rung that projects those recurrent axons downward. That response will vary, for one and the same sensory input, as a function of the prior and ever-changing 'cognitive context' (supplied by activation in the higher population of neurons that project the relevant recurrent pathways) to which that sensory input happens to arrive. That is, the response will now be a function of the ever-changing dynamical 'frame of mind' of the perceiving network.

This structural/functional addition to our basic feed-forward neuronal machinery brings with it a wealth of novel and welcome behaviors. For starters, a recurrent network can have its synaptic weights configured (i.e., it can be trained) so that the trajectories within its activation space will tend, as a function of where they begin, to spiral quickly in toward one or another of a family of distinct *attractor points*. These points can perform the function, roughly, that the familiar prototype points served in the case of the feed-forward networks we have already examined. But in recurrent networks those special activation points can be reached after several swift cycles of recurrent activity, cycles that *repeatedly* exploit the background information embodied in the network's synaptic connections. This often adds a further dimension of recognitional focus to the welcome phenomenon of vector completion that we have already noted, with approval, in the behavior of purely feed-forward networks. If the second rung of a recurrent network, given a slightly ambiguous or degraded sensory input, fails to zero in on a learned prototype on its first pass, by sheer feed-forward vector completion alone, it has the opportunity to 'settle in' on that prototype after several cycles of activation within the recurrent loop that feeds back to that rung. A prototype activation that failed to occur on the first hit from the input rung can nevertheless find a bull's-eye realization after several cycles of recurrent assistance. A recurrent network can thus be one notch wiser than its purely feed-forward analogue, even where single points in activation space are concerned.

Well and good. But it is the recurrent network's expanded capacity for *representation* that must command our primary attention. And the main enhancement they offer is their capacity to learn, not just a family of prototype *points* in the activation space of a given rung, but prototypical *trajectories* in the activation space of that rung. This allows them to represent prototypical sequences of events in the external world. It permits them to acquire a structured library of prototypical trajectories in activation space, trajectories that represent prototypical *causal processes* in the world, just as the simpler feed-forward networks acquire prototype points that represent prototypical *physical structures* in the world. This grip on causal processes unfolding in time is their principal virtue, for this alone allows any creature to navigate a figurative ocean of interacting causal processes. Without recurrent pathways and the dynamical virtues that come with them, any creature must be both blind to the causal processes that configure the universe, and powerless in the face of them.

For this reason, one should not think of recurrent axonal pathways as a subsequent articulation upon the more basic mechanism of a feed-forward network. The latter is more basic only in the incidental expository structure of these early chapters. Feed-forward networks are more easily grasped, and their simpler behaviors are more readily explored and explained. And so we examined them first. But recurrent networks are the rule in biological creatures, for they are essential to even the most rudimentary of biological functions. Consider the slow pulsing of the walls of an alimentary canal to force nutrients to traverse its length, the ceaseless beating of a heart to circulate blood, the grinding of a stomach to process food, or the regular contraction of lungs or gills to expel carbon dioxide and to ingest oxygen. All of these functions, and more, require executive neuronal control by a proprietary dynamical system that admits of stable limit cycles in its internal behavior—that is, a system that admits of cyclical trajectories in its neuronal activation space. Only if a network is suitably recurrent in its physical architecture can it display this essential dynamical signature. And so, evolutionarily speaking, recurrent networks are as old as nervous systems themselves—indeed, as old as animals themselves. For they are a nonoptional technology.

Not surprisingly, then, neuronal control of the basic biological functions just listed typically resides in the relevant biological organ itself, or in the spinal cord, or in the primitive brainstem of all vertebrates. It lies in what tradition calls the *autonomic* nervous system, because those near-ceaseless regulatory activities typically occur without ken or conscious connivance on the part of any creature. By contrast, the neuronal production and control of learned, occasional, deliberate, and conscious motor behavior typically resides in evolutionarily more recent brain structures such as the motor and premotor areas of the cerebral cortex. To be sure, these more recent areas embody 'know-how' as well, but it is knowledge of how to dodge an incoming projectile, how to throw a rock, how to swing a hammer, how to catch a fish, or how to start a fire, as opposed to knowledge of how to breathe, how to pump blood, or how to digest a meal.

In conjunction with the several sensory areas of the cortex (visual, tactile, auditory, and so on), these newer areas permit the creature that possesses them to engage and interact with the surrounding causal environment in much more intricate, insightful, and situation-specific ways. Such creatures can both see into a variety of possible futures (perhaps by *fast-forwarding* their learned perceptual prototype-trajectories, as and when current perception activates them), and then, in causal response to such off-line explorations, engage in informed motor behavior that brings about one of those possible futures at the expense of the evident and less palatable alternative futures.

In this way does creature *autonomy* slowly emerge from the universal causal order. Not because autonomous creatures manage somehow to *escape* membership in the all-inclusive causal order. Rather, it is because they have gained a sufficiently detailed conceptual grasp of that causal order's lasting background structure, and a perceptual grasp of its current configuration, to allow them to avoid, on a systematic basis, at least some of its less welcome prospective predations. Freedom, on this account, lies in the capacity to "dodge incoming projectiles," so to speak. Accordingly, freedom is not an all-or-nothing affair. To the contrary, it comes in degrees. One's freedom is jointly measured by how far and how accurately one can see into one's potential futures, and by how skilled one is at acting so as to favor (*causally* favor) some of those possibilities over others. Freedom, on this view, is ultimately a matter of knowledge—knowledge sufficient to see at least some distance into one's possible futures at any given time, and knowledge of how to behave so as to realize, or to enhance the likelihood of, some of those alternatives at the expense of the others. Freedom, perhaps ironically, consists not in evading the general causal order. Rather, it consists in having a detailed cognitive grip on the general causal order, one that exceeds the cognitive grip of barnacles and oysters, a grip that permits and sustains a complexity of interaction with the surrounding causal order that is forever denied to such sessile, passive, and pitiful hostages to fate.

Withal, we humans remain a part of the causal order, no less than the barnacles and the oysters. In our own and vastly more intricate ways, we too are presumably hostages to fate, if in ways that only an omniscient and supernatural deity could possibly comprehend. (Such a deity would need a keen sense of tragedy.) But that is not the point. The point is that we are not sessile, uncomprehending, and passive, in the fashion of the barnacles and oysters. We are very different from those creatures— as a matter of evident and incontestable empirical fact—for we can, and we do, dodge an endless variety of "incoming projectiles" in the face of which those lesser creatures are both blind and powerless. We can recognize, and decline to ingest, a poisonous mushroom. We can espy a marauding predator, and take suitable cover. We can see a social conflict developing, and spread oil on troubled waters. All of these cognitive skills

reflect our acquired moral character and our strictly unpredictable evaluative activities. Barnacles have no such skills. *That* difference is both real
and substantial. The measure of that difference is the measure of human
freedom.

Note finally that human freedom, from this perspective, is something
that can and does increase with advancing knowledge, and is strictly relative to the causal domains thus commanded. As a child's anticipatory and
practical skills develop, so does her freedom, for so does the modest control
she can exercise over her own fate. And as humanity's capacity for anticipating and shaping our economic, medical, industrial, and ecological
futures slowly expands, so does our collective freedom as a society. That
capacity, note well, resides in our collective scientific knowledge and in
our well-informed legislative and executive institutions. By contrast,
whereof we are wholly ignorant, thereof we are truly blind and powerless
hostages to fate. Thereof we are indeed mere driftwood, carried along by
the raging river, devoid of any hint of freedom. The cunning trout beneath
the river's surface has a significant degree of freedom; the piece of driftwood has none.[5]

Simply being embedded in a universal causal order is thus not inimical
to freedom. Indeed, being thoroughly embedded therein is the first requirement if real freedom is to be possible. Instead, it is the complete and utter
lack of control over one's unfolding history that is inimical to one's freedom.[6]
To the degree that one gains the missing knowledge, and deploys the
control that such foresight makes possible, to that degree has one gained
and exercised a measure of freedom. Freedom consists in the capacity for
the foresightful manipulation of one's own future. To be free is to hold at
least some (alas, only some) aspects of your own fate in your own hands.
Beyond this, there is no more valuable form of freedom to which we might
aspire. And the recurrent architecture of one's neural networks, to repeat
the point of this discussion, is what makes such freedom possible in the
first place. For it is our recurrent networks that simultaneously provide a

5. Dan Dennett (2003) has recently articulated, in *Freedom Evolves*, a novel compati-
bilist view of freedom along these same general lines, which I commend to the
reader's attention. What intrigues me is how his independently compelling view, or
something very like it, emerges so naturally from the overall account of causal
cognition and motor control held out to us by the theory of dynamically active
recurrent neural networks. This welcome "consilience of inductions" is, as we shall
see, only one of many in the area.
6. The neurobiology of self-control is briefly explored by Suhler and Churchland
(2009).

measure of insight into the future, and a measure of motor control over that future. Thus is freedom born, if only incrementally.

But we are not yet done with issues from traditional metaphysics. Putting motor control aside for a minute, let us ask: What is it we learn when we come to grasp a typical causal process in the world, such as the production of thunder by lightning, the transfer of motion when one billiard ball strikes a second, or the production of heat by fire? What is it that our learned trajectories in neuronal activation space really represent? Here the tradition lies heavy upon us. Accounts of causation, and of our knowledge thereof, range from David Hume's rather deflationary view that learning causal relations is merely learning 'constant conjunctions' between (instances of types of) events in the physical world, to David Armstrong's view that learning causal relations is coming to appreciate the existence of contingent but entirely real experience-transcendent relations between the relevant universals themselves, as opposed to chronic spatiotemporal relations between their many instances.

Despite their different levels of willingness to trade in Platonic currency (Armstrong, yes; Hume, no), both views display a common mistake, or so I wish to suggest. The shared mistake is the presumption that distinct types of particulars—whether spatiotemporal or Platonic—are grasped first, and then the relevant 'cause-indicating' connection between them is somehow divined afterward, as a contingent addition to an otherwise unconnected situation. In both cases, the relevant types of particulars are presented as being both epistemologically and ontologically primary, while whatever nomic or causal relations that might unite them are presented as being both epistemologically and ontologically secondary.

We need to consider the possibility that this gets things exactly backward. Given the overwhelming presence of recurrent pathways throughout any creature's brain, and given its primary need to get an immediate grip on the causal processes relevant to it—such as sucking milk from a mother, getting warm in the crook of her arm, recognizing the typical gait of one's conspecifics, recognizing the play behavior of one's siblings (and the often fierce competition for mother's resources that they provide)—it is at least worth considering the hypothesis that prototypical *causal* or *nomological processes* are what the brain grasps first and best. The creation and fine-tuning of useful *trajectories* in activation space may be the primary obligation of our most basic mechanisms of learning. The subsequent recognition of distinct types of *particulars* is something that may depend for its integrity on a prior appreciation of the world's causal structure, and, as we have seen, that appreciation requires the development of a structured family of prototype trajectories. It requires the development of an activation-space map of the intrinsic structure of the *four*-dimensional spacetime world. It requires that the brain acquire a grasp of *four-dimensional* universals. It requires that the brain develop a systematic representation of typical four-dimensional structures instanced again and again throughout the four-dimensional manifold of universal spacetime.

Those of us who already have a firm grip on the world's 4-D structure will naturally have a difficult time appreciating, at first, the infinite variety of alternative possible ways in which one might partition the 3-D world into distinct 'objects.' Consider the perverse, but a priori acceptable, 'object' that consists of the left one-fifth of my computer's keyboard, the right half of the crumpled manila envelope partly underneath it, and the blue letters "Dell Dimension" printed just above both on the lower border of my CRT monitor. We would not normally count this as a particular thing because it displays no unitary pattern of causal interaction with the rest of the world, or even among its internal parts. Pull on one part of it (the envelope, say), perhaps the simplest test for a unitary physical object, and the rest of this tripartite 'object' will not even come along for the ride. By contrast, real physical particulars of one type or another are distinguished as such, and one type is distinguished from other types, by the characteristic weave of 4-D causal processes in which they are embedded as a sequence of more-or-less invariant 3-D temporal slices.

A baseball-sized rock, for example, rolls as a whole when bowled along the ground, flies as a whole when thrown, sinks as a whole when placed in water, shatters a pane of glass when thrown against it, crushes a walnut shell when used as a hammer, and so on at considerable length. An underwater trout displays an equally long but quite different suite of causal behaviors, as does a tumbleweed, a chisel, or a V-shaped flock of migrating geese. Each counts as a distinct type of particular, not just because each is extended in time (after all, save for point events, *everything* is extended in time, including the perverse 'object' scouted above), but because each is a relevantly invariant participant in a distinct family of prototypical causal processes.

Suppose we agree that what elevates a candidate particular (or a candidate kind of particular) to the status of a *genuine* particular (or to the status of a *natural kind* of particular) is the characteristic family of genuine causal processes in which it chronically participates. We still have to ask, what is it that elevates any unfolding process to the status of a *genuine causal process*, as opposed to a mere accidental sequence or a pseudo-process, such as the motion and successive shape-changes of my shadow along an irregular wall as I walk its length, or the 'belted homerun ball' as portrayed on the luminous surface of my TV screen? (These are both cases, note, where the successive events at issue are *not* causally related to their respective temporal predecessors, on the wall or on the screen, but rather to a background *common cause*—to my moving body occluding the Sun in the first case, or to the real events in the real ballpark in the second—that gives rise to each secondary sequence.) And whatever the differentiating feature is, ontologically, how does the brain manage to divine it, epistemologically? In short, what makes a genuine causal process, and how do we detect it?

One possible answer is that a genuine causal process is an instance of a genuine four-dimensional universal, which is in turn a reflection of a truly exceptionless or invariant behavioral profile displayed (usually quite widely, but not always) within the objective structure of the four-dimensional continuum that embeds everything. This view of causal necessity goes back, past Hume and Armstrong, to the ancient Stoic logician Diodorus Cronus, who drew the distinction between mere contingency and genuine necessity solely in terms of time: the necessary features of the world are those that are genuinely time invariant; the contingent features are those that display variation over time (see Mates 1961).

This account originally struck me as entirely too crude and simplistic when I first encountered it, some forty-plus years ago. It might capture mere nomic necessity, I thought, but it obviously fails to discriminate that weakest form of necessity from more stringent forms such as analytic or logical necessity. And in any case, it faces the worry of mere accidental uniformities. Throughout the entire history of the universe (runs one example), there might never exist a cube of solid gold more than ten miles on a side. The unrestricted, nonvacuous, universal conditional statement, "All cubes of solid gold are less than ten miles on a side," would therefore be true. But that fact would surely not confer any nomic necessity or causal significance on that merely *de facto* 'regularity,' nor would it supply any *explanatory* clout with regard to the whys and wherefores of any existing gold cube with dimensions well short of ten miles.

Since that time, however, my confidence in my objections to Diodorus has faded. Thanks primarily to Quine, the analytic–synthetic distinction has deservedly become an embarrassing historical curio, as has the wholly mythical flavor of necessity that went with it. And thanks to quantum mechanics (and thanks also to the no less humbling example of Euclidean geometry confronting General Relativity), our so-called geometrical, arithmetical, logical or formal truths are starting to look like merely natural laws or nomic truths, if and where they are still true. Nomic necessity has reemerged as the gold standard of necessity in general. Diodorus would approve.

Even the worry about accidental regularities has lost much of its bite, for the universal nonexistence of a ten-mile cube of gold invites, and admits of, an entirely cogent *alternative* explanation in terms of the *initial conditions* that happen to characterize the universe, rather than an explanation in terms of the invariant *causal processes* that characterize it. By contrast, a plutonium cube, of the same dimensions, is indeed causally impossible, for reasons we can explain. Thanks to considerations of critical mass regarding neutron liberation, an accreting plutonium cube would explode long before reaching the specified dimensions. But gold need suffer no such dimensional limitations, as our knowledge of many of the world's *other* regularities allows us to infer (plutonium nuclei do, but gold nuclei do not, constantly leak disruptive neutrons).

There is no circularity here in appealing to the existence of other nomic regularities, already grasped, in order to distinguish the accidental regularities from the genuinely causal regularities. We make the same sort of appeal in denying causal status to the flowing shadow and the unfolding TV-screen images. Specifically, we have a much *better* account of why those pseudo-processes behave as they do, an account that locates the genuinely causal mechanisms at work somewhere else: in my Sun-occluding mobile body, and in the events in the ballpark, respectively. Just as with truth in general, nomic truth can be reliably determined only by a process that seeks consistency, systematicity, and simplicity within a broad web of other presumed truths. The real causal processes—the genuine four-dimensional invariants—form an interlocking and mutually reflecting whole. Even a partial grasp of that whole will allow us to identify mere accidental uniformities as such (e.g., the absence of gold cubes larger than ten miles on a side), without ever bothering to try to actually construct one so as to refute the relevant universal conditional directly. That de facto 'uniformity' can be left standing, like a trillion other 'uniformities' of equal irrelevance, without compromising the distinction between accidental and lawlike uniformities.

This point is reinforced by the observation that *any* universe with a genuine nomic/causal structure must also, and willy-nilly, display indefinitely many (absurdly many) purely accidental uniformities of the uninteresting kind at issue, save for that singularly rare universe in which absolutely *every* nomic possibility (such as our ten-mile gold cube) is eventually realized, somewhere and somewhen. Such a universe would have to be infinite in both space and time, in order to accommodate every possible temporal sequence of events, and every possible sequence of such sequences, and so on. Short of that singular case, however, every nomically possible universe must inevitably sport endless accidental uniformities of some sort or other, above and beyond what is strictly dictated by its real causal structure. But they can and will be unmasked as such, by cognitive creatures within that universe, by using the systematic procedure touted above.

3   Concept Formation via Hebbian Learning: Spatial Structures

Falsely seeing nomic/causal structure where there isn't any to be seen is only one of the dangers that a learning brain must try to avoid. *Failing* to see nomic/causal structure when it *is* there, but is buried in local noise or other confounding factors, is a no less salient danger. This returns us to the scientific (as opposed to the metaphysical) question: How does the brain manage to *generate* (roughly) accurate (partial) maps of the universe's four-dimensional structure? What is the process by which such maps of possible-causal-processes are actually constructed?

We have known the answer to this question for some time, but we understand it only poorly. In biological creatures, the process of experi-ence-dependent long-term adjustment of the brain's synaptic connections is definitely not governed by the supervised back-propagation-of-errors technique widely used to train up our computer-modeled artificial net-works. That brute-force artificial technique requires that the 'correct behav-ior' for a mature network be known in advance of any learning activity, in order that subsequent synaptic changes can be steered by the explicit goal of reducing the degree of error that separates the student network's actual behavior from the optimal behavior that this supervised technique seeks, stepwise, to impose upon it. But biological creatures have no such advance access to "the right answers," and they have no means of applying such information, to each synapse one-by-one, in any case. Synaptic change in biological creatures is apparently driven, instead, by a process called *Hebbian learning*, in honor of the psychologist Donald O. Hebb, who first described the process (see Hebb 1949).

The opening essentials are as follows. Think of a single neuron whose cell body is contacted by perhaps 750 excitatory synaptic connections from various other neurons, connections made with the tips of the many axonal end-branches reaching out from those other neurons. Let us suppose that the various strengths or 'weights' of those synaptic connections are dis-tributed randomly, at the outset, before any activity-dependent learning takes place. Now let those many other neurons begin to send sundry messages, in the form of occasional voltage-spike trains of various frequencies, along their many axons, toward our waiting student neuron. (Those incoming axonal messages, we shall assume, reflect the ongoing perceptual experience of the creature involved.)

Hebb's claim was that, whenever a high-frequency spike train, arriving to a given synaptic connection, finds our postsynaptic student neuron already in, or already rising to, a state of high activation, this temporal *coincidence* of high activation-levels has the subsequent effect of slightly *increasing* the strength or weight of the particular synapse involved. Accordingly, if a given synapse is *regularly* the site of such a temporal coincidence of arriving-excitatory-message and already-excited receiving neuron, its intrinsic strength or weight will expand substantially. That synapse will slowly but surely gain an 'increased level of influence or control' over the ongoing activation-levels of our student neuron. By contrast, those synaptic connections that chronically *fail* to find such temporal coincidences between arriving and local excitation-levels will have their weights go unenhanced (they may even be slowly reduced—this is called "anti-Hebbian learning"). Those synapses will thus have their level of influence over our student neuron gradually extinguished, at least relative to their more successful neighboring synapses.

This apparently pointless process of synaptic enhancement begins to make a little more sense when one appreciates that the most likely reason *why* the receiving neuron might already be in an elevated state of excitation is that a substantial number of its 749 *other* synaptic connections (a third of them, or 250, let us suppose) are *also* bringing excitatory messages to that very neuron at the very same time. The Hebbian process of synaptic enhancement thus favors whichever *cadre* of synaptic connections chronically tend to bring their occasional spike trains to the receiving neuron *simultaneously*. Roughly, whichever subset of synapses happen to 'sing' *together*, when and if they do sing, subsequently have their individual 'voices' made permanently louder, and for that very reason.

This special treatment reflects the fact that, whatever diverse and idiosyncratic sensory information happens to be carried, on that occasion, by each of the 250 synapses in this peculiar cadre, each bit of information concerns something sensory that *happened at the same time as* all of the other sensory events responsible for the activity in all of the other synapses in the cadre. If that simultaneous pattern of diverse sensory inputs is a pattern that is regularly *repeated* in the creature's ongoing experience, the subsequent activity of our receiving neuron will thus be made progressively more sensitive to that pattern in particular, at the expense of its potential sensitivities to other possible patterns.

It is worth noting that one and the same neuron can become, in this way, selectively sensitive to *more than one* quite distinct sensory input pattern. Distinct cadres of synapses—mostly outside of the 250 supposed in our initial example—can form their own time-united cadres within the larger population of 750, cadres whose voices are similarly enhanced by progressive Hebbian amplification, cadres that make the receiving neuron selectively sensitive to a second, a third, and a fourth possible sensory input pattern. A given synapse may even enjoy membership in more than one such preferred cadre, and may thus participate (at distinct times) in signaling the presence of distinct sensory input patterns. A neuron, in sum, can have more than one 'preferred stimulus.'

The graphic illustration of figure 3.7 will fix the basic process of Hebbian learning firmly in one's imagination. Behold an artificial 25 × 30 'retinal' population of sensory input cells, every one of which projects a proprietary axon that makes a weakly excitatory synaptic connection to the single receiving cell to the right of the retina. (Such a receiving cell would typically be part of a large population of second-layer cells, but let us focus for a moment on just one cell therein.) Let us suppose that these 750 synaptic connections have various, randomly set positive values not far above a synaptic strength of zero.

We are now going to present, to this population of figurative retinal cells, a series of images such as the various 25 × 30 patterns displayed underneath the rudimentary network of figure 3.7. All of them, let us suppose, are random distributions of roughly 50 percent black squares scattered among 50 percent white/gray squares, except one, the image at the lower left. Plainly (to you and me), it represents a human face.

To the network, however, it is distinguished in no relevant way from the others, for it, too, is composed of roughly 50 percent black squares distributed among 50 percent white squares. But let us suppose that, during the training period that we now initiate, this uniquely face-patterned image is presented to the network ten times as often as any of the other images, whose less frequent presentations are distributed randomly throughout this instruction period. If the 750 synaptic connections with the second-layer cell are governed by the Hebbian rule described several paragraphs ago, then the peculiar cadre of 375 connections (roughly half of the total of 750) projecting from the 375 white/gray (i.e., somewhat illuminated) squares of the face-patterned image will have a clear advantage over all of the other connections arriving thereto from the black

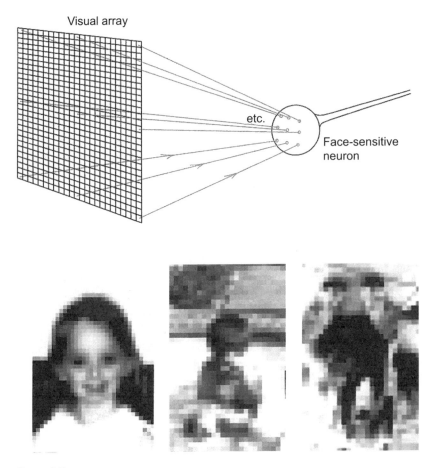

Figure 3.7
Making a neuron preferentially sensitive to a specific input.

squares. Since they are collectively activated ten times more often than any other cadre, the strength or weight of each connection in that favored cadre will grow ten times more quickly than the elements of any alternative cadre. After extensive training, therefore, they will all be at least ten times stronger than any connection outside that cadre.

The upshot of this period of synaptic-weight adjustment is a configuration of synaptic weights, at the target or second-layer cell, that makes that cell respond, to any presentation of the face-image in figure 3.7, with ten times the activation or excitation level that is produced by any other of the retinal inputs encountered in the training set. Thanks to its preferentially strengthened cadre of synapses, the face-pattern has become its 'preferred stimulus.' Moreover, and as with population-coding networks in general, any subsequent input pattern that roughly *resembles* that training face-pattern will *also* produce a preferentially high (though not a maximum) level of activation in the receiving cell. That is to say, the network is also preferentially sensitive to things *similar* to that face image.

This Hebbian process of preferential synaptic enhancement, and the sheer statistics of the network's perceptual experience during its training period, have transformed our utterly naïve network into a primitive 'face-detector.' Nor was any 'supervision' by a knowledgeable instructor required. No 'evaluations' of the network's 'performance' are made at any time, let alone subsequently used to shape the network. Nor is there anything special about the face image at the lower left, *save* for the fact that this pattern happened to dominate the network's unfolding perceptual experience. Had any alternative pattern shown such domination, then *that* pattern would have slowly become the second-layer cell's preferred stimulus. Hebbian-trained networks are thus fiercely 'empirical' in their learning habits: it is the *frequently repeated* patterns of raw perceptual input that gradually come to dominate the network's highly selective neuronal responses within the layers above the input layer.

The example here deployed is visual in its cognitive achievement, but a similar story would reveal itself if the input population had modeled, instead, the hair-cells of the cochlea, and if the instantaneous sensory patterns presented to that distinct input population were acoustic in character. Such a network could be similarly trained, by nothing but the spatiotemporal statistics of its own experience, to react preferentially to a specific *voice*, if that voice were to present itself much more frequently than any other auditory input. Or, if the input patterns reflected sensory activity in the olfactory system, or in the gustatory system, it could be trained to respond selectively to certain *smells*, or to certain *tastes*. The Hebbian learning process itself is indifferent to, and wholly ignorant of, the distal and objective reality that happens to drive any network's sensory activity. All it cares about is the repeating activational patterns, if any, displayed within that ongoing activity.

Two major cautions must be noted immediately. Our rudimentary network, even after successful training on images such as those posted above, is still a long way from having the concept of an individual face, as an enduring physical object. For that, much more is needed, both in the character of the training data and in the physical organization of the network needed to exploit it, as we shall see in the pages to come. Second, our network is biologically unrealistic insofar as it 'completely connected':

that is, *every* retinal cell projects an axon to, and makes a synaptic connection with, the receiving cell at the second layer. Real brains are nowhere near so profligate in their connectivity. Nor could they be. There is simply not enough *room* on the receiving surface (its dendritic tree included) of a typical neuron to find moorage for each of the 500,000 axons arriving up the optic nerve from the ganglion cells at the retina. A single cell might plausibly receive synaptic connections from perhaps a thousand arriving axons (thanks largely to the broad arbor of its branching dendrites), conceivably even two or three thousand of them. But half a million is too much.

Accordingly, each cell in the LGN is doomed to receive something under 1 percent of the total information arriving to the LGN from the retina, in contrast to the 100 percent displayed by the receiving cell in the artificial network of figure 3.7. Real neural networks, that is, are typically only *sparsely* connected. This is almost certainly a good thing, for reasons that are quickly made obvious. If the network of figure 3.7 were given, let's say, 100 receiving neurons at the second layer (rather than just one), all of them completely connected to the retina in the same way, then all of those cells would end up with exactly the same 'preferred stimulus.' They would all behave exactly as did their single-cell precursor. This would serve no purpose beyond an insane redundancy.

If, however, each of those 100 cells receive only *a random 1 percent* of the 750 projections from the unchanged retinal layer, then Hebbean learning on the training set displayed in figure 3.7 (with the face-pattern once again enjoying a tenfold advantage in its frequency of presentation) will now produce a considerable *variety* of preferred stimuli across those receiving neurons. After training, each one will continue to display a measurable fondness for our single face-image, to be sure. But since each second-layer cell receives a different and rather small (1 percent) 'sample' of that image, the acquired preferred stimulus of each receiving cell will produce a weaker response, it will likely be idiosyncratic to that cell, and it will include a broader range of distinct input stimuli than did the preferred stimulus of our solitary receiving cell of the original network. (This is because diverse *non*face input patterns are more likely to *share*, with the face image, the specific 1 percent—7 or 8—of its pixels that project to any given second-layer cell.)

This apparent 'degradation' of its original cognitive capacity can be a very useful thing. For if we now repeatedly present a *variety* of quite different face images to the enhanced network, beyond the single face image in figure 3.7, each such image will have a somewhat *different* effect on each of the 100 receiving cells. Each face image—the old one and the new ones—will thus produce a positive but idiosyncratic *profile* of activation levels across the 100 receiving cells. Any face image will produce at least some response across all 100 cells, but since their individual 'preferred stimuli' are no longer identical, each cell will react with a different level of activation. Overall, the result is that any particular face image will thus produce a signature *profile* of activation levels at the second layer, a profile that is unique to that individual face image. These diverse second-layer profiles will thus be diagnostic of which particular face image has been presented to the retina on any given post-learning occasion. They provide a 100-dimensional *assay* of any input face, and they produce a unique assay for each of the different faces it was trained on.

Each of those 100 second-layer cells is of course only very broadly tuned to faces. That is to say, a considerable variety of distinct faces will each produce *some* level of response in a given cell, if only a feeble response. Taken collectively, however, those 100 distinct, broadly tuned, but mutually overlapping sensitivities allow the network-as-a-whole to make some very *narrowly* tuned responses to specific faces. It can 'triangulate'—or more impressively, it can literally *centulate*—the exact location of its current target in a 100-dimensional face space.

This opens up the possibility of now adding a *third* layer to our Hebbian-trained network (see fig. 3.8), a layer that can be trained to discriminate between the faces of distinct individuals, if those several individuals are

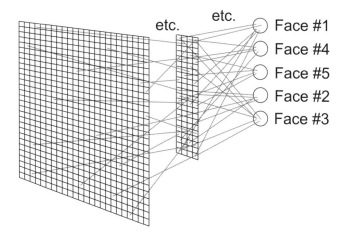

Figure 3.8
Preferential sensitivity to a special *range* of stimuli.

all frequently encountered in the network's training experience. This is because the profiles of activation at the new 100-cell second layer are, again, just more patterns of activation, the preferentially repeated occurrence of which will drive the Hebbean process of making the cells yet another rung higher in the processing hierarchy preferentially sensitive to whatever patterns dominate the collective behavior of whatever neurons happen to provide *their* input. With our new third layer, however, the input patterns to it are not purely sensory patterns. They are patterns of activation in a neuronal population already one step removed from the sensory periphery.

We have already examined, in section 5 of chapter 2, a network that successfully exploits such second-layer activation-profiles, namely, Cottrell's face-recognition network. That network was trained not by Hebbian updating, but by supervised learning: specifically, by back-propagation. Evidently, however, such a training regime is not the only training procedure that can yield a closely similar result. Let us therefore add a third layer to our already enhanced network—say, five cells above the second layer, each one of which is sparsely and randomly connected to perhaps 20 percent of that second population. And then let us provide this expanded network with a new training regime in which *five* distinct face-images dominate the network's otherwise noisy sensory experience at the first population in the hierarchy. This will produce, by the Hebbian process discussed above, five distinct (but frequently produced) face-profiles or face-assays that dominate the activation patterns at the second layer, profiles to which each of the third-layer neurons will eventually become selectively sensitive. Indirectly, then, each of those five cells at our new third layer is quite likely to acquire one of those five face images as its preferred stimuli, and quite probably (given their sparse and random connections to layer two) each will focus on a face different from any of the other four.

We now have a network that can discriminate any one of the face images of our five distinct individuals by displaying a distinct activation-profile across its five-cell third layer. It will not perform as accurately in this regard as Cottrell's network, but it will display the same sort of cognitive capacity. And it achieves this capacity with no supervision by a prescient agent. The abstract information-sensitivity of the Hebbian learning process, the connectivity of the network, and the objective statistics of its sensory inputs will do the job all by themselves.

What is perhaps most important about this kind of learning process, beyond its being biologically realistic right down to the synaptic level of physiological activity, is that *it does not require a conceptual framework already in place*, a framework fit for expressing propositions, some of which serve as hypotheses about the world, and some of which serve as evidence for or against those hypotheses. As remarked earlier, the standard philosophical accounts of the learning process—inductivism, hypothetico-deductivism, Bayesian probabilistic coherentism—all presuppose a background framework of meaningful concepts already in play, a framework within which their preferred forms of proposition-evaluation must take place. Where such conceptual frameworks come from to begin with— that is, how they originate—is a matter that is either ignored or badly fumbled by all of those accounts. (Recall our earlier critiques of Locke/ Hume-style Concatenative Empiricism on the one hand, and Descartes/ Fodor-style Concept Nativism on the other.) The Hebbian story explored above—concerning how the raw statistics of a creature's sensory experience can make the creature differentially sensitive to certain prototypical patterns of neuronal activation—offers an account of how a structured *family* of prototypical categories can slowly emerge and form the background conceptual framework in which subsequent sensory inputs are preferentially interpreted. It offers an account of learning that needs no antecedent conceptual framework in order to get an opening grip on the world. On this account, our concepts emerge gradually as an integral part of our growing understanding of the objective, lawlike relations that unite them and divide them.

All of this, however, is preamble. What we still need to address is how Hebbian learning can subserve the learning of specifically *temporal* structures. To this vital matter we now turn.

4 Concept Formation via Hebbian Learning: The Special Case of Temporal Structures

The progressive Hebbian enhancement of any synaptic connection is a function, as we saw, of the temporal conjunction of an excitatory signal arriving to a specific synapse at the same time that the postsynaptic neuron is already in a state of raised excitation. This temporal coincidence, of pre- and postsynaptic excitations on both sides of that site, ramps up the weight of that synaptic connection, especially if that coincidence is repeated frequently. The favored synapse subsequently has a larger effect on the receiving neuron, for the same strength of arriving axonal message, than it did earlier in its excitatory career.

This profile of selective signal augmentation can be deployed, in a network with a recurrent architecture (see fig. 3.9a), to selectively 'reward'

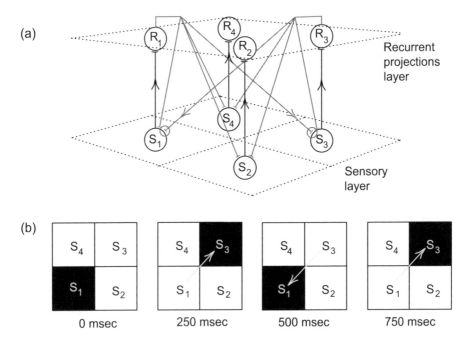

Figure 3.9
A maximally simple sequence-learning network.

exactly those synaptic connections that happen to make accurate *'anticipa-tions'* or *'prediction'* about the enhanced excitational state of the postsyn-aptic neurons to which they happen to connect. The two-layer arrangement of the network in figure 3.9a will illustrate the point. Note first that, col-lectively, the feed-forward or upward-reaching connections are all topo-graphic in nature. That is to say, the *left-of, right-of,* and *betweenness* relations, that structure the lowest or sensory layer of neurons, are all preserved as we ascend this (minimally brief) hierarchy of subsequent neuronal populations. All of the feed-forward axons point 'straight up,' and they all make a strong or high-weight excitatory connection with their receiving neuron.

By contrast, the recurrent or descending pathways (here portrayed in gray) from any given upper-rung neuron are all strongly 'radiative': they all branch so as to project back to each and every one of the neurons at the immediately preceding layer, where they make weak but identical excitatory connections. (I here portray the descending or recurrent pro-jections of only *two* of the four upper-layer neurons—R₁ and R₃—to avoid clutter in the diagram.) The point of this 'topographic-up' and 'radiative-down' arrangement begins to emerge when we consider a *sequence* of stimulations at the sensory layer, as portrayed in the strip of 'movie film' in figure 3.9b, a sequence that might crudely represent, for example, the ongoing behavior of a swinging pendulum as seen from above.

In the first frame of the movie, only S_1 is activated by sensory input. In the second frame, 250 msec later, only S_3 is activated. In the third frame, at $t = 500$ msec, only S_1 is activated. And so on, back toward S_3 again, at intervals of 250 msec, so that the locus of sensory activation slides steadily back and forth with a cycle time of 500 msec, that is, with a frequency of 2 Hz. This constitutes the sensory rung's (very rudimentary) ongoing representation of the unfolding behavior of a swinging pendulum. We may think of those four square 'neurons' as being the light-sensitive elements of a rudimentary 2×2 'retina.'

Let us finally suppose that the time it takes for any one-step axonal message to be conducted to its receiving neuron is one-eighth of a second, or 125 msec. Let this be true for descending messages as well as for ascending ones. If this is so, then you will note that 250 msec after the S_1 sensory neuron has been activated, the second-layer cell directly above it, R_1, will already have sent four excitatory signals back downward, to each and every one of the neurons at the sensory layer. But at the time those descending signals reach those four sensory neurons—at 250 msec—only neuron S_3 is being activated by the then-current sensory input to that layer (see again fig. 3.9b, the frame at 250 msec). Accordingly, only the descending synaptic connection (circled in gray) from R_1 to S_3 will be the beneficiary of a Hebbian boost in its synaptic weight on that occasion. The three sibling connections elsewhere in the sensory layer, by contrast, receive no such Hebbian encouragement, because their target sensory neurons—S_1, S_2, and S_4—are all quiescent at 250 msec. On this occasion at least, the timing of those other descending messages is wrong. They do not arrive to an already excited neuron. And so, the weights that convey those messages are not increased.

Note that exactly the same sort of process will be reprised 250 msec after neuron S_3 gets activated by our oscillating external input. For when activated, it too sends an ascending signal that activates the second-rung cell directly above it, R_3, which then sends a branching axonal signal back down to all four of the sensory neurons. However, only the R_3-to-S_1 branch will have its signal arrive to an already-activated sensory neuron, S_1, which at that time (500 msec) is already registering the bottom-left extreme of the pendulum's swing. In this second case, and after a total of 500 msec has passed since the pendulum's cycle began, it is the R_3-to-S_1 descending connection (circled in gray) that receives a selective Hebbian boost in its synaptic weight.

In two stages of 250 msec each, the oscillatory sensory activity of our network has produced a small Hebbian boost in the weights of exactly *two* descending synaptic connections, circled here in gray. Our first pendulum-swing cycle is now complete, and exactly two synapses (out of a total of sixteen, not all shown to avoid clutter) have had their weights selectively, if only slightly, enhanced.

It is both helpful, and relevant, to see the branching descending signals from the second-rung cells as *predictions* concerning which of the sensory neurons will *next* (i.e., in 125 msec) be active. For it is precisely the *successful* predictions that yield an *increase* in the participating synapse's weight. The failed predictions yield no such increase. Indeed, if we suppose that an *anti*-Hebbian learning process is also at work here (a process by which chronic *failures* of excitatory coincidences yield a progressive *reduction* of a synapse's weight), then those *un*successful connections will shrink in weight or even disappear entirely. In any case, the *relative* influence of those predictively unsuccessful connections will shrink in the face of chronic presentation of the swinging pendulum pattern to the student network.

Accordingly, the network's behavior will increasingly come to be dominated by the predictively successful connections, for those are the ones whose weights have grown larger. If the oscillating sensory sequence depicted in figure 3.9b dominates the experience of the network, the two predictively successful weights will slowly increase to rival the initially strong, and quite unchanging, feed-forward weights. The network's behavior, in the face of arbitrary sensory input, will then approximate the network of figure 3.10, wherein the predictively unsuccessful connections have simply disappeared, and the successful ones now have excitatory weights comparable to their feed-forward cousins.

Our much-instructed network now has the interesting feature that it shows a positive tendency to generate, all on its own, the same cycle of activity initially produced only by external sensory stimulation of the full pendulum-pattern at the sensory layer. It has become a self-sustaining oscillator that will produce, at the sensory layer (and at the second layer, too, for that matter, although it will there be 125 msec delayed), the periodic swinging pattern at issue, and it will do so even with partial or degraded versions of that sensory input to start it off. Once the descending

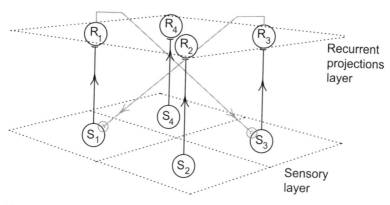

Figure 3.10
The dominant recurrent projections after training.

connections have grown their weights to values comparable to the weights of the original feed-forward connections, a strong stimulation of the S_1 sensory neuron, for example, is sufficient to start off the entire sequence, as is a strong stimulation of only S_3.

As well, the occasional masking of the oscillatory sensory input to either of these two sensory neurons will not undermine the network's cyclical behavior, once initiated. Our original network, thanks to its extended experience, has become selectively sensitive to a specific temporal pattern of sensory events, namely, that produced by a pendulum swinging on a diagonal path with a period of 2 Hz. And it will now engage in the temporal analogue of what network modelers have long called *vector completion* (recall the performance of Cottrell's face network in fig. 2.14). Specifically, our network, after long training on the specific pattern at issue, will engage in the completion of a prototypical vector *sequence*, of a prototypical *trajectory* in activation space, even when the sensory input representing those sequences is somehow degraded or partially masked.

Indeed, once the relevant temporal sequence has been learned, specifically *sensory* stimulation is strictly unnecessary. A single strong stimulation at the *second*-layer neuron R_1 (or at R_3) will also kick the quiescent network into the activation cycle at issue, illustrating the possibility that our temporally extended, learned oscillatory representation can also be activated by events above and beyond real sensory inputs from the external world. Our network, that is, is also capable of *imaginative* activity. It is capable of representational activity that is prompted not by a sensory encounter with an instance of the objective reality therein represented, but rather by 'top-down' stimulation of some kind from elsewhere within a larger encompassing network or 'brain.'

Unlike much larger networks of the same general configuration (specifically, uniformly topographic projections forward, and uniformly radiative projections back down), this tiny network is limited to learning only two or perhaps three prototypical sensory sequences at a time. But even this tiny network is pluripotent, to borrow an expression from cell biology. That is, it is capable of learning, from the sheer statistics of its own experience, a range of distinct temporal sequences beyond the specific example just explored. An oscillation between S_2 and S_4 is equally learnable, of course. So is an oscillation between S_4 and S_3, or between S_1 and S_2, or both jointly, as in a unitary vertical bar oscillating left and right across the sensory surface. Each of these can be acquired as the uniquely 'preferred stimulus' for the entire network, by once again exploiting the compulsive 'predictions' made by all of the recurrent projections, or, more accurately, by exploiting the selective Hebbian enhancement of those lucky projections whose predictions happen to be successful.

5 A Slightly More Realistic Case

Given the extreme simplicity of this network, its repertoire of possible learned trajectories is of course quite small. But we can easily see how to scale it up to encompass a rather larger range of more complex possible sequences. Instead of a 2×2 array of sensory neurons, let us contrive a network that has sixteen sensory neurons, arranged on a 4×4 surface, projecting straight up to a layer of sixteen second-layer neurons, as portrayed in figure 3.11a. And let each one of those second-layer neurons project back to each and every one of the sensory neurons, to make an initially weak excitatory recurrent connection, just as before. (I here illustrate the recurrent pathways from only *one* second-layer neuron, to avoid clutter and increase clarity. But let us assume that every neuron at the second layer makes a similar spread of downward-leading projections.)

Given its larger sensory surface, the range of input pattern sequences possible for the network is evidently now rather more extensive. But the principle displayed in our original 2×2 network—namely, the selective 'rewarding' of those recurrent connections that happen to make successful 'predictions' concerning which sensory neurons will be highly activated at the next temporal stage of the network's unfolding experience—is just

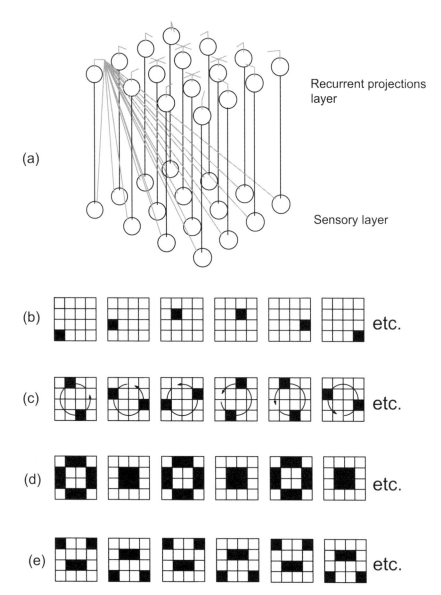

Figure 3.11
A somewhat larger sequence-learning network.

as relentlessly operative here as it was in the simpler case. This larger
network, too, will become selectively sensitive to whatever temporal patterns happen to dominate its unfolding sensory experience.[7]

Four possible examples lie underneath figure 3.11a. (Once again, I use
the movie-clip mode of presentation.) Figure 3.11b illustrates a parabolic
trajectory, as might represent the path of a thrown projectile. Figure 3.11c
illustrates a roughly circular pattern of sequential excitations of the sixteen
sensory neurons, as might represent the two painted tips of a propeller
slowly rotating at 0.5 Hz. Figure 3.11d illustrates a crude 'mouth' quickly
opening and closing at about 2 Hz. And figure 3.11e illustrates (very
crudely) the flapping-wing motion of a bird as seen from head-on, also at
2 Hz. These, too, will all be learnable by repeated presentations at the
sensory layer. That is to say, the network will become increasingly effective
at generating such a familiar sequence of activations, even in the face of
partial or degraded sensory inputs. The network as a whole will acquire
any such pattern as its 'preferred stimulus,' but this time the patterns are
spatio*temporal* rather than merely spatial, as with purely feed-forward networks. Our network here is capable of getting a firm grip on structures in
time as well as structures in space.[8]

The network is limited, of course, in the range of cyclical frequencies
to which it can become sensitive, whatever their spatiotemporal pattern.
Since the delay time of any neuron-to-neuron signal was stipulated to

7. This network architecture and learning procedure was modeled (in MatLab),
by Mark Churchland (Neurobiology Lab, Stanford), with a neuronal population of
100 at each layer. A small family of utterly random activation-sequences was independently generated and then fed to the network as training inputs. The network
performed very well at learning these target patterns, even with noise added to both
the training inputs and to the subsequent sensory inputs used for testing.

8. The recurrent character of this Predictor Network will remind some readers of
the classical Hopfield Net (see Hopfield 1982). But we are here looking at an importantly different creature. A trained Hopfield Net will go through many cycles, given
an initial (degraded) sensory input, in order to settle or relax into a previously
learned *spatial* pattern. Its cycling behavior in time is a means to a purely spatial
end. By contrast, the trained Predictor Net here portrayed responds immediately, to
appropriate sensory input, with a *temporal* pattern, which may happen to be oscillatory, but generally will not be. The specific recurrent activity it displays is not the
means to some end; it is the end itself. And there need be no relaxation involved
at all. The Predictor Net does, however, share an analogue of one of the Hopfield
Net's classical problems: the problem of overlapping target patterns. More on this
presently.

be 125 msec, no cyclical pattern with a frequency higher than 2 Hz can possibly hope to gain an accurate representation by the student network. (Although some *illusions* will be possible here. If the periodic input trajectory has a frequency of $n \times 2$ Hz, where n = integer 2, 3, 4, etc., then the network may well respond to it with some oscillatory trajectory of 2 Hz, since it will be 'picking up on' every second or third or fourth element in the input sequence, while being blind to the intervening elements. Think, for a possibly real example, of looking up at the whirling blades of a large military helicopter overhead. If the rotational frequency is right, you can *seem* to see the individual blades moving, rather slowly. But of course, you are not. The blades are moving much too fast for that.)

The value of 125 msec—the interneural delay time—was stipulated for purely illustrative purposes, of course. I wanted the artificial network to be able to learn a pendular sequence of 2 Hz. In humans and animals the typical delay time of interneuron communication is much shorter—closer to 10 msec—although it varies as a function of interneuronal distance, axon myelinization, and other factors. Given the specific axonal distances involved in our primary sensory pathways, this figure suggests that humans will have an absolute frequency limit, for the oscillatory sensory inputs they can grasp, of about twelve times that of our artificial network, or a limit of $12 \times 2 = 24$ Hz. But we are nowhere near that good. A flying sparrow flaps its wings at about 8 Hz, and I can just barely see that process. So 24 Hz is asking a lot. But these values are (very roughly) in the right ball-park. Oscillations at or beyond these frequencies are quite impossible for us to see. We will revisit this issue.

How the network manages to activate its learned sequences, even with only partial or degraded sensory inputs, is not at all difficult to appreciate. Figure 3.12a will make the process transparent. I here present the network of figure 3.11a with all of its descending or recurrent pathways erased, except for that small subset that were favored by extensive Hebbian train-ing on the circling-square pattern of the sequence in 3.11b. And I here portray those experience-favored neuronal pathways in a sequence of increasingly dark grayscale lines, to indicate the temporal sequence of their real-time activation. (For simplicity's sake, I here portray the activation-sequence for only one of the propeller's two tips.) The analogy of a sequence of dominoes stood on end, and in a circle, comes to mind. Tip over one, and the rest will follow suit. And if, after toppling, each domino is restored to its vertical position (i.e., the relevant neuron is repolarized), the overall circling pattern of activation will repeat itself indefinitely.

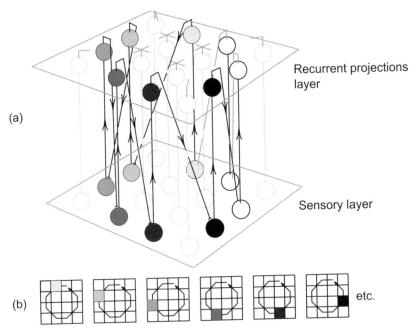

(a)

Recurrent projections
layer

Sensory layer

(b)

etc.

Figure 3.12
An unfolding sequence of activations.

6 In Search of Still Greater Realism

It also worth pointing out that the gross *microanatomy* of our several primary sensory pathways—visual, auditory, and somatosensory—all reflect the features central to the model of figure 3.11a, with one major exception. What is there portrayed as a 'sensory layer' is not a sensory layer at all, not in us, nor in other mammals. Rather, that layer is one step *removed* from a genuinely sensory layer that projects topographically upward, to be sure, but receives no descending projections at all. We are here talking about the mammalian retina. To get an anatomically more realistic artificial network, we need to add a true sensory layer underneath the arrangement already discussed, as is finally portrayed in figure 3.13.

This arrangement is still a considerable oversimplification, since there is a total of six major neuronal layers within the visual cortex itself, layers unacknowledged in the topmost population of this diagram. In fact, the LGN's axonal projections upward make their connections only to the neurons in layer 4 of the visual cortex (V1), which neurons then project back to layer 6. And it is the layer 6 neurons from V1 that finally project

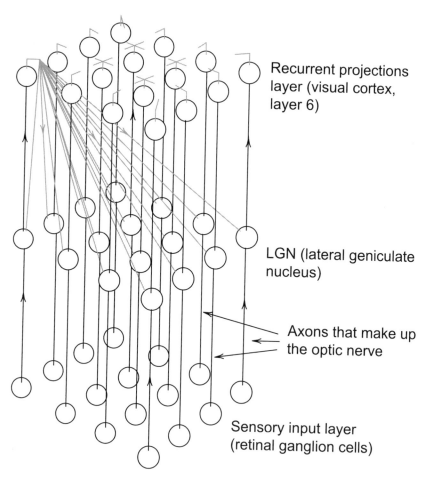

Recurrent projections
layer (visual cortex,
layer 6)

LGN (lateral geniculate
nucleus)

Axons that make up
the optic nerve

Sensory input layer
(retinal ganglion cells)

Figure 3.13
Cartoon depiction of the primary visual pathway in humans.

their decidedly more radiative projections back down to the LGN (see Sherman 2005).

As well, those recurrent projections do not all go directly back to the LGN neurons. A substantial fraction of them synapse onto an intervening population of smallish *inhibitory* interneurons (located in the thalamic reticular nucleus), which serve as 'polarity inverters' before finally synapsing—now with an inhibitory connection—onto sundry neurons in the LGN proper. This wrinkle is no doubt a useful one. The existence of such inhibitory connections means that the activated neurons in visual cortex will also, and willy-nilly, be making a 'spray' of predictions about which

LGN neurons will *not* be active in the next spatial pattern of sensory input, in addition to their more direct excitatory predictions about which sensory neurons *will* be active. If synapses that chronically bring an inhibitory message to an already quiescent neuron are also selectively enhanced, in a distinct version of Hebbian learning, then the network will also be learning what typically *won't* happen as a prototypical causal sequence unfolds, as well as what typically will happen. Such a complementary process will sharpen the network's cognitive resolution and help to suppress the occasional noise that degrades the network's sensory inputs.

In particular, this inhibitory wrinkle on our original, purely excitatory, network will serve to ameliorate the problem (alluded to in note 8 of this chapter) of distinct learned sequences that share a common or overlapping spatial stage. This is a problem because when that shared spatial pattern is activated, by whatever input, it will then tend to complete or set in motion *both* of the temporal sequences that happen to share that sequence element, even though only one of them accurately reflects the network's ongoing sensory input. If the mutual overlap of that shared spatial pattern is absolutely perfect, this problem is unsolvable, at least with the specific neuronal architecture here on display.

However, if one of these two (overlapping) spatial patterns is merely a proper *subset* of the other, then its subsequent excitatory influences will tend to be canceled out by the collective inhibitory influences of the larger number of neurons that register the dominant or superset spatial pattern. The more robust sequence-element will thus tend to suppress the less robust sequence-element, an asymmetric suppression that will quickly be amplified as the dominant sequence unfolds. The inhibitory interneurons thus appear to play an important functional role: they may help to solve the problem of partially overlapping spatial patterns, at least in large networks.

As well, the problem arises in the first place only because we required that the connectivity of the network's recurrent pathways be *complete*, that each upper-layer neuron project a connection back to each and every one of the lower-layer neurons. But this idealization is biologically unrealizable in a network with many thousands, or, more likely, millions of neurons. A real neuron can make little more than a thousand or so synaptic connections with other neurons. Accordingly, we must concede that, in a realistic network, a given upper-layer neuron can boast recurrent connections to only a smallish percentage of the preceding population.

But as those connections get sparser, our spurious-stimulations problem gets progressively less likely. This is because the probability of a single neuron's correctly predicting *two distinct phases* of the very same periodic process falls swiftly as overall connectivity is reduced to more realistic levels. Accordingly, the problem of perfectly overlapping stimulation-patterns, and the doubled-up sequences that they produce, appears to be an artifact of the hyper-simple models with which we began. The problem, which was minor to begin with, fades steadily as the networks approach realistic numbers and plausible profiles of connectivity.

But we are not yet done adding wrinkles, for the LGN has a small handful of distinct neuronal layers of its own, unlike the single sheet portrayed here. All of this will help to account for the fact that the human limit on graspable frequencies is only about half of the 24 Hz of my initial, oversimplified, cartoon-based guess. Our own neuronal system contains rather more axonal paths to traverse and synaptic connections to bridge than appear in the schematic cartoon of figure 3.13. Our own recurrent system has perhaps four distinct axonal steps, each taking perhaps 10 msec to traverse, for a total of about 40 msec for a single cycle up-and-back. Our perceptual frequency limits should thus be correspondingly lower— 1000 msec/(2×40) msec = around 12 Hz—as indeed they are. That is to say, an appropriately corrected model predicts, apparently correctly, the rough limits of human visual perception where the discrimination of oscillatory frequencies is concerned.

Beyond this, the cartoon remains a recognizable portrait of the gross organization of the Primary Visual Pathway in mammals. In particular, the actual 'upward' projections from retina to LGN, and from LGN on to visual cortex, are indeed all pretty ruthlessly topographic, whereas the recurrent or 'downward' projections from cortex to LGN are strongly radiative and much, much more numerous (see again Sherman 2005, 115). In fact, the recurrent projections from the visual cortex back down to the LGN are roughly *ten times* as numerous as the feed-forward projections from the LGN to the cortex. (Perhaps surprisingly, only 7 percent of the inputs to the LGN come from the retina. Instead, fully 93 percent of them come from higher neuronal populations such as—most prominently but not exclusively—the primary visual cortex! See Sherman 2005, 107.) Why are those descending connections there at all? And why are they there in such nonspecific radiative profusion?

The trajectory-learning story sketched above offers a possible explanation of this curious arrangement. One of its functions (it may have several) is that it makes possible the learning of prototypical causal sequences in the observable world. Here again, our optimism must be tempered: there is a large gap, still to be bridged, between learning prototypical temporal sequences and learning genuine *causal* sequences. The latter are a proper subset of the former and creatures are chronically vulnerable to seeing the latter in the former, especially in cases where the creature has no means of intervening in the relevant flow of events, thereby to explore their often hidden causal structure. On the other hand, if our learning here proceeds by the process described in the preceding pages, it is perhaps no surprise that we are as vulnerable to this sort of illusion as we famously are.

As we scale up the schematic organization of figure 3.13 so that each layer contains hundreds of thousands of neurons, the range of learnable temporal sequences increases dramatically, of course, as does their complexity. One and the same network can learn to recognize a wide variety of quite different prototypical sequences. A running basketball player will then become distinguishable from a skating hockey player, and both of these will be distinguishable from the lunging left–right gait of a cross-country skier. Even so, all three of these familiar trajectories will probably be close together in the all-up activation space of the network's middle layer—as distinct variations on a roughly common theme. Indeed, any individual hockey player's idiosyncratic skating gait will itself be represented by a highly complex activation-space trajectory whose many elements cluster fairly closely around the elements of a central, prototypical *skating* trajectory slowly carved within the relevant activation space by long experience with many different skaters. That is to say, the results of the Hebbian-induced temporal learning we are here discussing will also involve the construction of diverse focal *prototypes*, each surrounded by variant instances, and all of them located, at appropriate distances from one another, within a relevant and coherent *similarity space*. And that, let us remind ourselves, is what the learning of a conceptual framework amounts to, even in the case of temporal structures.

I close this discussion by emphasizing once more the decidedly non-Humean and preconceptual nature of the learning process here described. In particular, I want to emphasize the possibility—nay, the probability—that the learning of typical temporal sequences does not proceed by *first* grasping the one-, two-, or three-dimensional spatial configurations of typical physical objects and properties, and then *subsequently* noting their temporal conjunctions and progressive modifications over time. Rather, I suggest, what the brain picks up, first and best, are the typical *four-*dimensional sequences that it encounters in raw, unconceptualized experience. Typical kinds of physical objects, and their characteristic properties, are presumably something that the brain manages subsequently to divine therefrom, as lower-dimensional invariants that play a constituting role in those salient and repeatedly encountered four-dimensional sequences. Such objects and properties are a subsequent abstraction from, not the epistemological ground of, our understanding of the world's causal structure. For such understanding can be gleaned from the world, as we have seen in the preceding pages, without any prior conceptual grasp whatever of either typical objects or typical properties. From the point of view of the brain, then, it appears to be processes that are primary, not things and their properties.

This wholly unconscious, preconceptual, and subconceptual Hebbian learning process can yield, as we saw, conceptual frameworks for representing both prototypical spatial and prototypical temporal organizations, frameworks that can *correspond* quite accurately to at least some of the objective structures of various external feature-domains. Such correspondences can arise quite naturally, because those conceptual frameworks were causally *produced* by protracted causal interaction with those same objective structures. This claim is not the metaphysical assertion of some transcendental epistemology. Rather, it is the innocent and entirely fallible assertion of an empirical account of how brains function, an account that derives its current credibility, such as it is, from its explanatory and predictive successes in the domain of human and animal cognitive activity. How brains can generate reliable representations of (some of) the world's objective structure is no more problematic, in principle, than the question of how a *camera* can generate reliable representations of (some of) the world's objective structure. The brain is just a great deal better at it than is a standard optical camera. And so our theories of how the brain comes to represent the world will need to be much more sophisticated than our theories of how a store-bought camera comes to represent the world.

There is, to be sure, the reflexive twist that, in undertaking our epistemological theorizing, *we are the camera*, a camera attempting to represent its own capacity for world-representation. But there is no insurmountable mystery in our position here, any more than there is for an optical camera. For example, we can set up an open-sided box-camera with a simple lens at the center of one face, a plane of film at the opposite face, with an apple (say) some distance in front of the lens casting an upside-down image of an apple on the film's orthogonal surface. We can then glue in some taut white threads to indicate the paths of the relevant rays of light projecting from the apple, through the lens, and to the inverted image. And finally, we can use a second, identical camera to take a picture of all of this, as viewed from the side. The resulting picture, we can all agree, is an elementary metarepresentation of how a camera constructs a first-order representation of the objective apple. And this metarepresentation is created inside another camera. Elementary optics textbooks are full of precisely such pictures, or of diagrams isomorphic thereto. There is much about the picture-taking process, to be sure, that there goes *un*represented. But our reflexive representer here is only a camera. A brain can aspire to much greater epistemological achievements than this faint and feeble cousin.

7 Ascending from Several Egocentric Spaces to One Allocentric Space

A shortcoming in the account of representation—of both the enduring and the ephemeral—so far presented in this book is that none of these representations look past the idiosyncratic format of each distinct sensory modality, or the idiosyncratic spatial perspective of each distinct perceiving individual. But one's true epistemological situation is nowhere near so blinkered. We each live in the same unitary three-dimensional space, a space populated with interacting physical objects that last through time. And we know that we do. We may move through that world, but the bulk of the objects that it contains do not move at all, or only rarely. My kitchen sink, its surrounding cupboards, my brick fireplace, the streets and buildings outside, the rocks, and trees, and hills in the distance collectively form a stable if variegated grid, a grid within which I can learn to recognize my current spatial position, and through which I can generally move at will.

When I do move smoothly through that unmoving grid, I see, hear, and feel typical kinds of smooth changes. Most obviously, as I walk, drive, or fly forward I experience what visual scientists call "optical flow" (see fig. 3.14). Specifically, the fixed objects ahead of me grow steadily 'larger' and occupy positions in my visual field progressively farther away from its center, finally to disappear, with maximum velocity, at its periphery. Walking or riding backward produces exactly the opposite suite of 'behaviors' among visible objects: everything then flows toward the center of my visual field, shrinking as they go.

Gazing steadily sideways, at a right angle to my direction of motion, as when looking out the window of a train or bus, yields a different but related flow called "motion parallax," wherein distant objects seem barely to move at all, while progressively nearer objects flow by, with progressively higher 'velocities,' in the opposite direction. And if one is motionless, then rotating one's head left or right produces a simpler and more uniform flow of perceived objects, as does nodding one's head forward or backward.

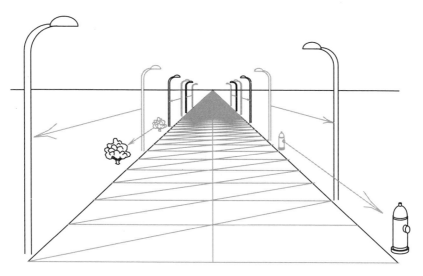

Figure 3.14
The "optical flow" typical of forward motion.

Of course, none of these 'behaviors' are genuine activities of the external
objects themselves. By hypothesis, those objects are motionless. Rather,
their 'apparent' behaviors are one and all the effects of a single *common
cause*, namely, one's own personal motion as one carves a trajectory among
them. But their collective 'behaviors,' as represented in one's several perceptual manifolds, give one a very firm cognitive grip both on their various
positions relative to each other, and on the nature of one's own motion
in and among them. Here, then, are the materials for the brain's construction of a unitary spacetime manifold, within which all objects are located,
and to which all motions and changes can be referred. The Kantian conception of a unitary spacetime framework as the "pure manifold of sensory
intuition" comes immediately to mind here, and the resources of the story
told in the preceding section are but a few steps away from reconstructing
it, as we shall now see.

The fact is, the topographic-feed-forward, radiative-recurrent sort of
network architecture explored in the preceding section is entirely capable
of learning, and subsequently of recognizing, precisely the sorts of chronically encountered optical flows just described. For as we saw, that architecture is capable of learning *any* sequence of unfolding activation patterns,
including utterly arbitrary ones, even in the face of noise. The comparatively simple and well-behaved sensory flows at issue are a piece of cake,
especially because they completely dominate the visual experiences of any

creature with the capacity for locomotion. You cannot turn your head, or walk across the room, without experiencing at least one of the several flows at issue. Your only way to avoid them is to remain absolutely frozen, as if you were a statue. Accordingly, we are all of us doomed to learn these flows, and to learn them very early in our cognitive lives.

As we do, we also divine that these optical flows come in exactly two distinct varieties. First, there are the *translational* flows, as when the body moves through the surrounding grid of objects. These are the flows in which the flow elements change their positions not only relative to our sensory manifolds, but also relative to each other within any such manifold. And second, there are the *rotational* flows, as when you rotate, in some way, your otherwise motionless head. These are the flows in which the flow elements rigidly maintain their within-manifold positions relative to each other, despite collectively sweeping across our sensory manifolds.

Within the class of translational flows in particular, we further learn that they divide into exactly three distinct subclasses. First, there are those that involve the apparent motion of a center-outward or a periphery-inward flow (in fact, a bodily motion forward or backward). Second, there are those that involve the always-graded apparent motion of a leftward or rightward flow (in fact, a bodily motion lateral to the direction of gaze). And third, there are those that involve the apparent motion of an upward or downward flow (in fact, a bodily motion vertically). Any possible translation flow will be some combination of these three basic kinds, a direct reflection of the fact that any possible trajectory of my body through the surrounding grid of motionless objects is some combination of translational motions in the three dimensions of objective space. In this way, and in others, the three-dimensionality of objective physical space is made accessible to the learning brain.

The very same lesson can be derived, independently, by a network's learning to distinguish between the three distinct subclasses of the quite different *rotational* flows, those flows in which the mutual positions of the flow elements in subjective space do not change relative to one another. First, the elements can all flow uniformly leftward or rightward (as when one turns one's head smoothly toward the right or left). Second, the elements can flow uniformly upward or downward (as when one nods one's head down or up). And third, the elements can all rotate around a locus at the center of one's visual field, as when one cocks one's head to the left or right. Any possible rotational flow will be some combination of these three basic kinds, a direct reflection of the fact that any possible objective rotation of my head is some combination of objective rotations around the three orthogonal axes of objective space. Thus, the two kinds of optical flow—translational and rotational—are jointly unanimous in their respective indications of the dimensionality of objective physical space. It is *three* dimensional, and one can either translate, or rotate, or both, along or around any and all of those three dimensions.

One's ongoing auditory experience displays a related suite of perceptual 'flows,' as when one moves through an unmoving population of various noise-emitting objects. (Think of riding a wheel chair, blindfolded, through a large cocktail party of noisy but unmoving interlocutors.) Here, to be sure, the flow-elements are not so exquisitely well focused as in the case of the optical flows. Once again, however, the translational and rotational auditory flows encountered both decompose nicely into the same trio of basic kinds, reflecting the same objective dimensionality of space revealed by the corresponding kinds of optical flows. Indeed, remove the blindfold and one can begin to assign a specific auditory-flow element (the baritone voice) and a specific optical-flow element (the beefy guy in the blue suit) to one and the same unmoving individual. That is to say, the optical and the auditory flows are both mutually coherent and mutually validating reflections of a single motion—one's own—through a single three-dimensional world of at least temporarily unmoving objects.

Such cross-modal integration does not end with vision and audition. One's vestibular system—the mutually orthogonal circular accelerometers inside the inner ear—supply the brain with continuously updated information concerning *changes* in the body's motion in physical space, that is, once again, changes in its objective translational and rotational activities. Not surprisingly, these representations accord nicely with whatever *changes* happen to be displayed in one or other of the concurrent optical flows (and indeed, the auditory flows) experienced by the same individual. As one turns from one translational heading to another (a twenty-degree turn to the right, for example), the radiant point of one's optical flow moves from being centered on one family of flow elements to being centered on a partially overlapping but now relevantly different family. One can both feel, and see, that one has changed direction slightly.

A further grip on exactly such changes is supplied by the body's somato-sensory and proprioceptive systems, for these also provide the brain with ongoing information about changes in one's motion. For example, pressing a car's accelerator pushes the driver's seat against his back, and the pressure is both obvious, and strictly proportional to the degree of acceleration involved. A right turn tends to make the driver's body tilt to the left, and the muscle-monitoring proprioceptive system advises the brain of the compensatory muscle contractions automatically applied by the motor system to keep the driver roughly upright through the turn. Such representations accord nicely with one another, and most importantly, with the vestibular, auditory, and above all, with the optical representations discussed earlier.

These diverse forms of representation embody systematic mutual information that includes *simultaneous onsets*, mutual conformity in the diverse ways in which they *evolve* over time, and even mutual conformity in the timing and nature of their several *changes* in the manner of their temporal evolution. Accordingly, if the information they severally contain could all be presented to a single two-stage population of neurons (of the kind explored in the preceding section) at which these diverse sensory pathways were physically to converge, then that population would be in a position to learn, by nothing more than Hebbian updating, a *unitary* representation of the world, a representation that is no longer hostage to the diverse and radically egocentric representations supplied by each of the peripheral sensory manifolds individually.

Where in the brain such a unitary and modality-independent representation might be found is a good question, to which I do not know the answer. The so-called 'primary visual cortex' is one candidate, as is the LGN itself, at least given the diverse and modally various inputs that arrive to each. But many brain areas meet this condition, and it may well be that there is no single Rome to which all roads lead. Cross-sensory integration may be a feature of the brain's processing at any stage and at every opportunity. An interacting coalition of brain areas might easily serve the unifying or consolidating functions contemplated here.

To pull off that Kantian trick, however, will require one further representational achievement, beyond simply coordinating diverse sensory inputs in a unified, modality-neutral way. Specifically, the representations must also abstract from, or factor out in some way, the inevitable sensory flux that causally derives from nothing other than one's own bodily motions, translational and/or rotational, through objective space. This subtraction might seem a simple enough task: just stand perfectly still—so as to bring all of those fluxes at least temporarily to zero—and then embrace the resulting representation as the canonical portrayal of independent reality.

But alas, the world looks and sounds *different* from each possible objective vantage point that you might visit, and different once again depending on which of the many ways you might happen to be facing when you do come to rest. Simply subtracting the sensory flow-information entirely—by bringing it to a halt—does little or nothing toward achieving our goal. We need to *use* that profoundly systematic flow-information, not lose it.

And use it we do. One learns to represent oneself as just one more physical object among the many that surround one, and one learns to represent much the greater part of one's sensory flux as a reflection of the motion of one's own body through an objective matrix of mostly unmoving other physical bodies. This can't be too hard, given the statistics of our terrestrial environment. After all, one's own motion is typically both swifter and far more frequent than the motion of most of the objects in a run-of-the-mill room, or a field, or a forest—which means that the optical and other sensory fluxes that one encounters are typically a reflection of one's own individual motion, rather than a reflection of wholesale motions among objects at large.

Moreover, as we saw several paragraphs ago, the several kinds of sensory flows actually produced in us—optical, auditory, and so on—all reduce to sundry combinations of just three basic and very well-behaved types. The appropriate dimensionality of the desired allocentric space is thus both moderately obvious to a learning network, and agreeably small: three. Representing oneself as a coherently moving object in an objective three-dimensional space brings almost everything else in this emerging perceptual universe to a screeching halt. Once brought thus 'to rest,' those other objects maintain a vast and mostly stable family of mutual spatial relations, a 'geographical' framework in which one can subsequently learn to remember specific spatial locations, and specific paths that reliably lead from one such location to another. One learns to find one's way around the mostly unmoving world.

Modeling exactly this achievement in an artificial neural network remains an unfulfilled ambition, at least by this author. But we have already seen the required sort of dimensionality-reduction displayed in several cases discussed so far: in Cottrell's face-recognition network, in Usui's color-coding network, in the gradually learned two-dimensional coordinative surface embedded in the eight-dimensional space of the crab network's middle-layer, and in the analytic techniques that successfully grasped the underlying dimensions of variation involved in the nodding and turning head. In the case contemplated in the preceding paragraph, the several dimensions of one's optical flows must first be identified as they unfold (as we saw, this is probably not a serious problem) and then 'discounted' or 'subtracted from' one's optical flows (this is the unsolved part of the problem) so as to leave a representation of a stable framework of motionless objects.[9] Against that stable framework, the subsequent learning of typical temporal sequences and presumptive causal processes can then proceed with substantially greater accuracy, since the changes due to one's own accidental perspective and occasional personal motions will have been subtracted from the all-up sensory flux that one encounters.

A final and maximally dramatic example of this sort of 'imposed' stability is the phenomenological stability of the objects on one's dinner table, for example, despite the many discontinuous saccades made by one's eyes as they dart from place to place several times each second during the course of dinner. The position of the elements in one's visual field change abruptly and repeatedly with each successive saccade, but one's subjective impression is nonetheless an impression of a stable and unchanging table-setting. This is because information concerning the timing, direction, and magnitude of each saccade is automatically sent from the ocular muscles back to the cortical areas that represent the visual world. Once again, and despite our philosophical tradition, it is the brain's antecedent capacity to monitor *motions* and *changes* intelligibly that allows us to finally get a grip on *un*moving physical objects and their sundry *stable* properties, not the other way around.

9. There is an irony to our cognitive achievement here: it leaves us profoundly prejudiced against the Aristarchan/Copernican/Galilean claim that the entire Earth is itself in constant translational and rotational motion. (No wonder the widespread incomprehension and resistance initially encountered by these several visionaries!) One thus has to learn, at least for the accurate reckoning of local planetary motions, to temporarily 'reset' one's allocentric grid so that the 'fixed stars' assume the role of motionless objects. This is not easy, at least for perceptual apprehension, but it can be done, and the results can be exhilarating. See Churchland 1979, 31, 33.

4 Second-Level Learning: Dynamical Changes in the Brain and Domain-Shifted Redeployments of Existing Concepts

1 The Achievement of Explanatory Understanding

One of the virtues of the map-indexing story of perceptual activity, as we saw, is its account of 'vector completion'—of the strong tendency that a trained network has to activate a vector somewhere within one of its acquired prototype regions in preferential response to a fairly wide range of sensory inputs, especially to partial, degraded, or somewhat novel versions of the typical sensory inputs on which it has been successfully trained. This reflects the trained network's automatic tendency to interpret any arbitrary stimulus as being, in some way or other, yet another instance of one-or-other of the enduring categories or causal profiles it has already discovered in the universe of its past perceptual experience. On occasion, this chronic impulse can positively mislead the network, as we have already seen. (Recall the false identifications, made by Cottrell's face-network, of the identity of completely novel individuals: they were frequently, and quite wrongly, assimilated to the individual in the original training set whose image they most closely resembled.) But on the whole, such a cognitive strategy serves the network well enough. In the real world, no two sensory situations will be strictly identical, so any effective network has no option but to engage in the assimilative strategy at issue. The occasional and inevitable mistakes must simply be endured, and perhaps learned from.

Such quasi-automatic assimilation, as we also saw, is often dramatically ampliative, in that the relevant input is assigned a specific position in cognitive space, a space that represents a prototypical *family* of *general* features and relations. As a result of such willful interpretation, the network is thereby made host to a range of specific *expectations* concerning features of the perceptual situation either unrepresented in its current sensory input (recall Mary's occluded eyes), or yet-to-be represented in its sensory input (as in the future-but-still-unfolding behavior of a typical causal process).

Such assimilation of a here-and-now 'part' to a background 'whole'—whether spatial, or temporal, or both—constitutes the network's *interpretation* or *understanding* of the situation currently under sensory apprehension. It constitutes the network's *explanatory take* on the phenomenon it confronts. And the adequacy, or inadequacy, of that explanatory construal will be measured by the accuracy of those multiple expectations as the network's subsequent experience reveals. Or, more strictly, it will be measured by the agreement, or disagreement, between those expectations and the network's subsequent sensory experience, as subsequently interpreted with the same background conceptual resources.

Such ampliative construals are the plebian stuff of one's minute-to-minute perceptual and practical life, as one's existing expertise deals effortlessly with the flowing familiar. But as experience expands, surprises occasionally show up, as when a child first encounters a Pekinese terrier, at rest in someone's lap, and takes a puzzled few minutes to assimilate the small beast to a natural kind—dogs—hitherto dominated by the examples of labs, golden retrievers, and German shepherds. In the same way, older children are surprised to learn that dolphins are not fish, but are aquatic mammals, distantly related to dogs rather than to salmon, in much the same way that seals and walruses are. Such surprising assimilations do pay predictive and explanatory dividends for the child. The pug-nosed Pekinese, however tiny and moplike, nonetheless barks, chases cats, and licks your face. And the dolphin, however fishlike in its appearance, is nonetheless warm blooded, gives birth to fully formed young, and must ascend regularly to the ocean's surface to breathe air. These predictive and explanatory dividends, and many others, of course, are the primary point of such novel assimilations. Expanding the extension of one's category *dog*, or *mammal*, in the specific ways described, gives one a deeper understanding of the novel things (the Pekinese, or the dolphin) thus reconstrued. They finally emerge, after due puzzlement, as unexpected instances of a category whose basic empirical profile is already familiar. They are brought within the ambit of one's antecedent and already functional understanding.

A reperception of this kind may involve some smallish cognitive effort, as the brain exploits its recurrent pathways to steer the problematic sensory stimulation, eventually, into the familiar cognitive harbor of the old but now redeployed prototype. Little effort is involved, perhaps, in the case of the Pekinese, especially if one is also shown a boxer, and then an English sheepdog. (Though these dogs are both quite large, the former has a pug nose, and the latter is dramatically moplike.) Such intermediate cases help the child bridge the similarity-gap that initially stumped him. Rather more strenuous cognitive effort is required in the case of the dolphin, no doubt, partly because dolphins are so ostentatiously fishlike, but also because most people have a rather feeble concept of *mammal* to begin with. Accordingly, for many folks, I suspect, dolphins remain as 'default fish.' Be that as it may, many of us have succeeded in making the relevant conceptual transition, and are thus better able to anticipate the behavioral and anatomical profile displayed by dolphins.

These two examples, modest though they are, are here proposed as entry-point philosophical models for conceptual redeployments that concern much more extensive empirical domains, and involve much more radical cases of conceptual change. A third example of what is basically the same sort of reassimilation, I suggest, is Darwin's reconstrual of the long history of plants and animals on Earth as a *natural* instance of the *artificial selection* long practiced by plant and animal breeders on the farm. The striking diversity of types of dogs, for example, was well known even then to be a reflection of centuries of selective breeding by human kennel-keepers seeking to maximize the frequency of occurrence, in subsequent generations of dogs, of some desired trait or other. The diversity within horses, cattle, grains, fruits, and vegetables provided further, if sometimes tenuous, examples of the same thing (history is long, and records are poor). Replace the approving/disapproving human breeder with an 'approving/ disapproving' natural reproductive environment, and replace the paltry history of human farms with the entire history of all life on Earth, and you have a rather startling reconception of the origin of all current species, and all past ones as well.

This reconception of the biological universe offered an immediate explanation of the ways in which anatomically specialized creatures are so well adapted to the highly diverse environments in which they live. It offered an immediate explanation of the geologically ancient fossil record and of the almost endless diversity of strange creatures that it contains. It made sense of the evident 'family tree' structure that slowly emerged from our reconstructing the details of that geological record. And it continues, even today, to make sense of the very same 'family tree' structure that emerges as well from our comparative analysis of the DNA molecules that make up the genomes of all currently living species. Indeed, the fact that those two, independently constructed trees systematically agree with one another—to the extent that we have access to their overlap—is presumptive testament to the integrity of both.

But for our purposes here, what is of interest about Darwin's presumed insight is that—plus or minus a deliberate human breeder—it consisted in the application of a complex concept (one that he already possessed) to a novel empirical domain. Before ever boarding the *Beagle*, he knew about the processes of *natural variation* and *selective reproduction* as mechanisms that could jointly produce uniformly novel types of creatures, and in large quantities of subsequently true-breeding types. And so did many others before him, the breeders themselves being foremost among them. What was special about Darwin was that he began to explore the possibility that those same mechanisms might be responsible, over geological time, for the totality and diversity of all species, our own included, that currently inhabit the Earth.[1] The operations of such a mechanism, he saw, need not be confined to the kennels and stables of human farms.

The subsequent details of that celebrated exploration need not concern us at this point. What needs emphasizing, for our purposes, is that the process was originally launched by Darwin's redeploying a fairly mundane concept, one typically deployed in the practical domain of the proximate here-and-now, in a new domain: specifically, the domain of deep historical time and the development of all life on Earth. A vast and deeply puzzling historical process was reconceived as an unexpectedly grand instance of a modest and already familiar causal process. This presumed insight did not involve any significant modification in the physical *structure* of Darwin's brain. In particular, it did not involve the deletion or addition of any synaptic connections therein, nor even any adjustments in the weight-configuration of his existing synaptic connections (not, at least, in the short term). Rather, the multistable *dynamical* system that was Darwin's brain was tipped by accidental circumstance into an importantly new regime of cognitive processing, at least where the topic of the Earth's biological history was concerned.

The causal origins of Darwin's explanatory epiphany resided in the peculiar modulations, of his normal perceptual and imaginative processes, induced by the novel contextual information brought to those processes via his descending or recurrent axonal pathways. That, at least, is what the neural-network approach on display in this book invites us to consider. A purely feed-forward network, once its synaptic weights have been fixed, is doomed to respond to the same sensory inputs with unchanging and uniquely appropriate cognitive outputs. Such a network embodies a

---

1. The mere suggestion of long-term evolution was not new: Spencer and Lamarck had earlier sketched such views. But Darwin's specific *mechanism* for the process was new. Only Wallace shared Darwin's specific insight here, and he was personally unable to accept it as the explanation of *human* origins in particular.

function—in the strict mathematical sense—and any function has a unique output for any given input. A feed-forward network, once trained, is thereby doomed to a kind of rigid cognitive monotony.

A trained network with a *recurrent* architecture, by contrast, is entirely capable of responding to one and the same sensory input in a variety of very different ways (see again fig. 3.6, plate 14). This pluripotency arises from the fact that the network's output vector on any occasion is determined by (1) the character of the relevant sensory input, as usual; (2) the fixed weight-configuration of its trained synapses, as usual; and most importantly, (3) *the current dynamical or activational state of every nonsensory neuron in the network*. For those higher-level neurons deliver a profile of excitatory and inhibitory messages back toward the very same neurons that receive the current sensory information. Most importantly, the dynamical factors comprehended under (3) can vary substantially, from moment to moment, and from situation to situation, even if the factors comprehended under (1) and (2) remain entirely constant. Diverse interpretations, construals, vector completions, or cognitive takes on any perceptual situation are thus made possible, even for a network with a fixed physical architecture. Though its background architecture remains fixed, its dynamical states can meander all over the map.

As those states meander, they provide an ever-changing cognitive context into which the same sensory subject-matter, on different occasions, is constrained to arrive. Mostly, those contextual variations make only a small and local difference in the brain's subsequent processing of that repeated sensory input. But occasionally they can make a large and lasting difference. Once Darwin had seen the now-famous diversity of finch-types specific to the *environmentally* diverse Galapagos Islands as being historically and causally analogous to the diversity of dog-types specific to the *selectionally* diverse dog-breeding kennels of Europe, he would never see or think of the overall diversity of biological forms in quite the same way again. And what gave Darwin's conceptual reinterpretation here the lasting impact that it had on him was precisely the extraordinary explanatory power that it provided, both in its breadth and in its depth. It offered a unitary account of the development of all life on Earth, throughout all history. It was also, note well, an account that conflicted sharply with the broadly accepted biblical account of the origin of species. It thus seized his attention, and shaped all of his subsequent conceptual activities on that topic, to a degree that few things ever could. The Platonic camera that was Darwin's brain had redeployed one of its existing 'cognitive lenses' so as to provide a systematically novel mode of conceptualization where issues of biological history were concerned.

The most important element in this particular account of Darwin's cognitive achievement is, of course, the top-down modulation, via his descending or recurrent axonal pathways, of his sensory interpretations, and indeed of his cognitive deployments in general. But only slightly less important is the claim that the specific conceptual resources thereby deployed are conceptual resources that he already possessed. They did not need to be laboriously created from scratch, by Hebbian learning, in protracted response to the empirical phenomena within the new domain. His antecedent conceptual resources, once happily repositioned, could hit the ground already running. We thus have a second mechanism that at least occasionally deserves the name *learning*, a mechanism over and above the ultimately basic mechanism of the gradual Hebbian modification of the brain's synaptic connections. This new mechanism is dependent, to be sure, on that prior and more basic learning process, for without it, the brain would have no concepts to redeploy in the first place. But it is a distinct mechanism, one distinguished by the comparative *quickness* with which a fortunate creature can gain a systematically new conceptual grip on some domain of phenomena. It can happen in a matter of seconds, instead of weeks, months, or years. We may call this more mercurial process *Second-Level Learning*, to distinguish it from its much slower precursor discussed at length in the preceding chapter.

The history of science is full of the occasionally useful epiphanies displayed in Second-Level Learning. Newton's sudden reperception/reconception of the Moon's familiar but highly puzzling elliptical trajectory (a reprise that supposedly occurred during the falling-apple incident at Woolsthorpe), as being just a large-scale instance of a flung stone's trajectory here on Earth, supplies a second major example of this strictly mundane assimilative phenomenon. Both trajectories were conic sections, after all: the flung stone follows a (roughly) parabolic path and the Moon an elliptical path. And Newton effectively displayed the many possible cases of the former as just so many partial instances of the latter with his striking diagram of the multiple launchings of progressively faster projectiles from an imaginary mountaintop (see fig. 4.1 for an updated version).

Each of the several suborbital projectiles hits the Earth after following the initial trajectory of an ellipse, an ellipse that intersects the Earth's surface because the projectile's original horizontal velocity is too low to avoid a collision with the Earth, a collision somewhere short of a complete

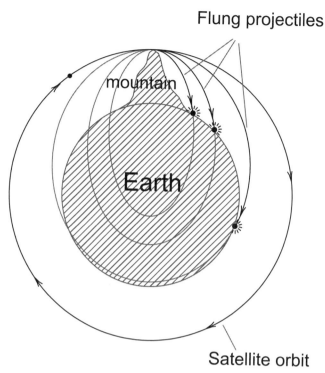

Figure 4.1
Isaac Newton's explanatory account of orbital motion.

cycle. But if the projectile is given a sufficiently high initial velocity, evidently its elliptical path must expand to surround the entire Earth, leaving the projectile in an endless loop that repeatedly grazes its launching point atop the mountain.[2] The Moon—dare we think?—is just such a projectile with a still larger radius of motion, a projectile whose constant acceleration toward the Earth is always 'compensated' by its roughly constant 'horizontal' motion at right angles to its ever-changing direction of fall.

This perspective is highly suggestive, since Newton could quickly and independently calculate, from the known radius of the Moon's orbit and its known 'horizontal' velocity, that its acceleration toward the Earth is 1/3600 of the known gravitational acceleration of an apple, stone, or any

---

2. This diagram ignores air resistance. As well, the lines inside the Earth represent what the projectile's subsurface path *would be* if all of the Earth's mass were concentrated in a point at its center.

other projectile here at the Earth's surface. This specific fraction—1/3600— is a most intriguing number, since it is precisely the *square* of 1/60, and the Moon was independently known to be 60 times farther away from the Earth's center than is our falling apple or flung stone here at the Earth's surface. (The Earth's radius is 4,000 miles. The Moon's distance is 240,000 miles. 240,000 ÷ 4,000 = 60.) If, like the flung stone, the Moon's earthward acceleration is due to, and directly proportional to, the force of the Earth's gravity, then that force apparently *falls off* as $1/R^2$, where R is the distance of the affected object from the Earth's center. In this way does Newton's novel explanatory take suggest a quantitative law concerning the force due to gravity, specifically: $F_G \propto 1/R^2$.

Nothing whatever guarantees that this law is true, but if it is true, then a host of other facts fall swiftly into cognitive place. For example, the uniquely massive Sun will simply have to be the center of all of the planetary motions, just as Copernicus had previously urged, on quite different grounds. Its gravitational force will dominate everything else's. And the Moon's orbit will have to be an ellipse, with the Earth at one focus. As of course it is. And as the Moon traverses that elliptical path, and speeds up toward its low point and slows down toward its high point, it will always carve out equal areas in equal times. As indeed it does. (These two regularities reflect Kepler's already familiar first and second laws of orbital motion.) And if the Sun's gravitational effect on all of its planets is also presumed to be governed by a $1/R^2$ law, then Newton can quickly deduce that the ever-longer orbital periods of its progressively more distant planets will have to obey a law of the form: orbital period $\propto \sqrt{R^3}$, where R represents the planet's orbital radius.[3] As indeed they all do. (This is Kepler's third law of planetary motion.) This latter explanatory insight was subsequently repeated in a further domain—namely, the relative periods of the four major satellites of Jupiter—on the assumption of a similar force law for Jupiter's gravity.

Nor did the insights end here. Newton's speculative ascription of mass to every body in the Solar System, and his ascription of a $1/R^2$ gravitational force centered on each, allowed him to see a large and disconnected set of further astronomical facts as the natural effects of the same underlying mechanism. Consider, for example, the roughly spherical shapes of the

---

3. Despite appearances, this insight is almost effortless. The seven-step deduction proceeds from three almost trivial premises beyond the presumptive gravitation law and consists of nothing but substituting sundry equals for equals and then collecting terms. See the appendix for the surprisingly brief details.

Earth, the Sun, and all of the planets. Consider the existence and timing of the Earth's oceanic tides. Consider the elliptical or hyperbolic paths of all comets. Consider the puzzling precession of the Moon's tilted orbit every 18 years. And consider the long-mysterious precession of the Earth's own rotational axis every 23,000 years. All of these diverse phenomena were successfully reconceived as just so many instances of forced motion due to a centripetal gravitational force that falls off as $1/R^2$. All of them.

These later conceptual accommodations took time, of course, time for Newton to explore his initial insight concerning the Moon's status as a stonelike projectile moving more or less uniformly under its own inertia horizontally, while accelerating toward the Earth vertically. Indeed, the full exploration of Newton's insight here consumed the energies of subsequent astronomical researchers for the next three centuries. The target domain steadily expanded to include the behavior of stellar systems far beyond our Solar System, the behavior of our Milky Way galaxy as a whole, and the behaviors, individual and collective, of all of the galaxies in the cosmos. Newton's dynamical redeployment of his existing concepts kindled a Kuhnian 'research program' of extraordinary depth and breadth, as has Darwin's, especially since the twentieth century's development of the field of molecular biology and our growing understanding of the structure and behavior of DNA. In the former case, we are not done even yet. (Recall, for example, the anomalously short rotation periods, recently discovered, of galaxies in general. Hence the suggestion of some unknown form of 'dark' matter, concentrated within each galaxy.) And in the latter case, to judge by the molecular intricacy and the broad range of the relevant metabolic and developmental phenomena, we have barely begun to put biochemical flesh on the macrolevel skeleton that Darwin bequeathed us.

The *dynamical redeployment* story on display in this chapter thus offers a natural account, not just of the neuronal dynamics of the occasional epiphanies themselves, but also of the *post*epiphany cognitive behaviors of the scientists who have been captured by them, both in their attempts to expand and consolidate the explanatory domains at issue, and in their sensitivity to the gradual emergence of stubborn anomalies, that is, to the emergence of salient phenomena that steadfastly *resist* assimilation to the prototypes or models under collective professional articulation. Kuhn was quite right to identify these long, postrevolutionary periods as consuming the intellectual lives of the great majority of scientists. He was right to identify the initial core of such research programs as being the Concrete Success, the identifiable example of an outstanding predictive/explanatory achievement. And he was right again to emphasize that expanding the reach of that favored paradigm is a process that is not governed by explicit rules and, indeed, is largely inarticulable in any case. From the perspective of dynamical learning, it is not hard to see why. The process involves the repeated attempt to apply a very high-dimensional concept—most of whose dimensions are inaccessible to its possessor—to a tentative range of perceptual or experimental phenomena, where (1) exactly *how* to apply the favored concept is somewhat unclear at the outset, and (2) exactly *what* experimental phenomena should be expected to yield to interpretation in its terms is similarly unclear. The process is thus doomed to be a groping meander in somewhat dim light, where both success and failure are often frustratingly ambiguous. Both 'confirmations' and 'refutations' are rarely, if ever, unequivocal.

2 On the Evaluation of Conceptual Frameworks: A Second Pass (Conceptual Redeployments)

Here is where issues of rational methodology begin to emerge. Both history and current experience show that humans are all too ready to interpret puzzling aspects of the world in various fabulous or irrelevant terms, and then rationalize away their subsequent predictive/manipulative failures, if those failures are noticed at all. This is not an on/off problem, but rather a graded threat that attends all of our cognitive activities. How best to minimize that threat, and maximize our grip on reality, is something we humans have been trying to learn for quite some time, and most especially since the several scientific advances of the Enlightenment, and the various philosophical commentaries that they inspired. If nothing else, scientific pioneers such as Galileo, Kepler, Descartes, Huygens, Newton, and Boyle gave us more and better *examples* of fruitful reconceptions of various empirical domains, namely, the various theories of the Early Modern scientific revolution.[4] This put us in a position to attempt to divine, perhaps, what epistemological features they all shared, and, correlatively, what features distinguished them from their less worthy conceptual precursors.

---

4. Recall, Galileo gave us the *moving ball that neither sped up nor slowed down*: a presumptive model for the eternal planetary motions. Kepler gave us the graceful *ellipse* as the actual path of planetary orbits. Descartes gave us the *fluid whirlpool* as the causal dynamics of the Copernican Solar System. Huygens gave us the *traveling wave* as the nature of light. Newton gave us the *flung stone* as the model for heavenly motion under the force of gravity. And Boyle gave us the *'spring' of air*.

We find inchoate early stabs in this vein from rationalists such as Descartes ("clear and distinct apprehension by the faculty of Reason"), and Leibniz ("accord with the Principle of Sufficient Reason"); and only slightly better stabs from empiricists such as Newton ("Hypotheses non fingo"), and Hume ("Matters of fact and real existence are known only by induction from experience"). But despite the presumed authority of their distinguished authors, these vague standards provide no real guidance for our cognitive behavior. Worse, they are positively *mis*leading. Trusting, with Descartes, to what *seems* clear to the current dictates of human Reason is a sure recipe for epistemological conservatism, as is restricting one's scientific generalizations to those reachable by Hume's process of Instantial Induction. This latter form of inference, note well, is also 'conceptually *conservative*' in the straightforward sense that it can yield no conclusions containing any descriptive predicates *not already contained* in the singular observation-sentences that serve as premises for such general conclusions. (Example: "This *crow* is *black*; that *crow* is *black*; those *crows* are *black*; therefore, all *crows* are *black*.") On this view, legitimate scientific laws must one and all be generalizations over strictly *observable* features of the world, such as crowness and blackness. How else to establish the premises from which such generalizations can be legitimately drawn? Unfortunately, the generalizations of modern science are mostly about humanly *un*observable regularities, regularities such as

The charge on any electron = 1.6×10^{-19} coulombs;
An electromagnetic wave is an ongoing interaction between mutually inducing electric and magnetic fields, oscillating at right angles to each other and 180° out of phase with each other, while propagating at 300,000 k/s in a vacuum;
A human DNA molecule is a long double-helical ladder with 3×10^9 nucleic-acid units as rungs;

and so forth, at great length. The connection between our general scientific knowledge and our singular experience is clearly much less direct than, and nowhere near so one-directional as, Newton and Hume naïvely surmised.

Subsequent philosophers in the Empiricist tradition got around this particular problem, with some real success, by appealing to the process of Hypothetico-Deduction instead of to the process of Instantial *In*duction. Karl Popper, for example, was quite happy to welcome laws and generalizations concerning perceptually *inaccessible* features into the realm of legitimate science, so long as those generalizations imposed, if only indirectly, systematic *constraints* on the perceivable or experimentally accessible world. This was Popper's famous attempt to demarcate the generalizations of Science from the comparatively disreputable generalizations of Metaphysics. Specifically, he required that a genuinely scientific hypothesis must logically entail, if only in conjunction with suitable auxiliary premises, the negation of at least some actual or potential observation-statements. This ensured, he thought, that the theoretical generalizations in question would be testable against experience. That is, they would be refutable by some possible experimental results. Anything worthy of the name "science," accordingly, would always be hostage to experience, if not in the simple-minded way previously outlined by Newton and Hume.

But Popper's story of the proper relation between science and experience was also too simple-minded. Formally speaking, we can always conjure up an 'auxiliary premise' that will put any lunatic metaphysical hypothesis into the required logical contact with a possible refuting observation statement. (For any such lunatic hypothesis "M," simply assume the auxiliary premise "if M then not-O," where "O" is some arbitrary observation statement. This will provide the required connection to a potentially refuting observation.) Accordingly, without some independent and non-question-begging criterion for what counts as a *scientifically legitimate* 'auxiliary premise,' Popper's proposed demarcation between Science and Metaphysics is a sham.

The proposal is a sham for a further reason. The supposedly possible *refutation* of a scientific hypothesis "H" at the hands of "if H then not-O" and "O" can be only as certain as one's confidence in the truth of "O." (The inference-rule *modus tollens* is perhaps as certain as one can get, but the refutation requires the presumed truth of "O" as well.) Unfortunately, given the theory-laden character of all concepts, and the contextual contingencies surrounding all perceptions, no observation statement is ever known with certainty, a point Popper himself acknowledges (see Popper 1972, 41–42, n.8). So no hypothesis, even a legitimately scientific one, can ever be refuted with certainty—not ever. One might shrug one's shoulders and acquiesce in this welcome consequence, resting content with the requirement that possible observations can at least *contradict* a genuinely scientific hypothesis, if not refute it with certainty.

But this won't draw the desired distinction either, even if we let go of the requirement of decisive refutability for general hypotheses. The problem is that presumptive metaphysics can also creep into our habits of *perceptual* judgment, as when an unquestioning devout sincerely avers, "I feel God's disapproval" or "I see God's happiness," when the rest of us would simply say, "I feel guilty" or "I see a glorious sunset." This possibility is not just a philosopher's a priori complaint: millions of religious people reflexively approach the *perceivable* world with precisely the sorts of metaphysical concepts just cited. In them, the high-dimensional conceptual map that gets vectorially indexed by their sensory systems represents the world quite differently from the conceptual map deployed by most of the rest of us. It is thus abstractly possible that their general metaphysical convictions, concerning, for example, what God does and does not demand from us, might be contextually contradicted by such 'observation judgments' as those just cited. Once again, Popper's proposal needs an independent and non-question-begging criterion of scientific character, this time for observation statements or perceptual judgments themselves. He cannot just help himself to a notion of what counts as observational. What we are—or are not—observing, measuring, monitoring, tracking, and perceiving is itself one of the questions that science must confront, and it must be prepared to evaluate a variety of competing possible answers.

This inescapable dependence on accepted background theory or prior conceptual commitments is entirely obvious when the question concerns the informational significance of the behavior of some artificial measuring instrument, such as an ammeter, or a thermometer. Is the needle-position of the ammeter measuring the volume flow of electric fluid? The number of electrons passing per time? The anger-level of daemons in the wire? Is the height of the thermometer's mercury column measuring the pressure of caloric fluid? The average kinetic energy of the ambient molecules? The outflow-rate of elemental phlogiston? We need to settle on some preferred theory or conceptual framework in order to answer such questions. In advance of such settled theory, unfortunately, measuring instruments do not come with their informational significance already written on their sleeves.

And neither do the various *biological* sense-modalities native to humans and other creatures, for these are no less measuring instruments than their artifactual brethren. Being part of the physical world, they respond, selectively and causally, to specific dimensions of objective reality. And they produce axonal outputs that eventually index a specific position in a humanly constructed, historically contingent, and quite-possibly faulty conceptual *map* of that objective reality. An endless variety of *different* conceptual maps are possible, for subsequent systematic indexing, by any given sensory modality in any given creature. Which map(s) we should embrace, as (provisionally) authoritative for the reflexive interpretation of our sensory experience, is part of the epistemological problem we face as cognitive creatures.

To be fair to Popper, he does confront the uncertainty of perceptual judgments, and he recommends that we be prepared, when necessary, to test them, too, against still further perceptual judgments and experimental explorations. This is surely good advice, and Popper deserves credit for giving it, but we have now lost any hope of articulating a decision procedure that will distinguish the merely metaphysical from the properly scientific. The touchstone epistemological level on which Popper's presumptive decision procedure clearly depends—the level of perceptual judgments—is evidently as subject to 'metaphysical' infection and corruption as is any other level.[5]

If we are thus denied any hope of authoritatively specifying what counts as a legitimately *scientific* (as opposed to a merely *metaphysical*) hypothesis, perhaps we might settle for something a little less ambitious and a little more realistic. Perhaps what is really important for scientific rationality is not the quasi-positivistic status of the *theories* that we entertain, but rather the programmatic character of the *methodology* employed to evaluate and to modify our hypotheses over time, often over very long periods of time. Popper himself, evidently aware of the problems with a straightforward falsifiability criterion for theories themselves, occasionally speaks, instead, of an ongoing "critical attitude"—as opposed to a "dogmatic attitude"— embraced by the *scientists* who propose and evaluate those theories. This shifts our philosophical attention away from individual theories and relocates it on the practice of the scientists, or better, on the communities of scientists, who deal with them.

Imre Lakatos, Popper's younger colleague and intellectual ally, constructed a detailed alternative to Popper's original focus on the evaluation of theories, an alternative where the important units of evaluation are *evolving research programs* instead of the various and fleeting theory-stages that such a temporally extended program might contain (see Lakatos 1970). On Lakatos's alternative account, competing research programs slowly reveal themselves, on the negative side, as stagnant or degenerating; or, on the positive side, as expanding and flourishing. The distinction resides, first, in the slowly discovered capacity or incapacity of the relevant research program to articulate its doctrinal core and/or expand its surrounding family of theoretical assumptions so as to increase the

5. The conceptual plasticity of human perception is explored at some length in chapter 2 of Churchland 1979.

presumptive 'empirical content' of the program as a whole. And second, it resides in the unfolding success or failure of that developing empirical content as it is successively confronted with the outcomes of the various empirical tests suggested by that research program.

On this alternative story (which, note well, remakes contact with Kuhn's sociological take on our scientific history), we both can and should tolerate our inability to specify, in advance, what counts as 'genuine empirical content,' since our judgments on that score may quite properly *vary* from research program to research program, and since the right to interpret the outcome of experiments, the behavior of our artificial measuring instruments, and even the true content or significance of our own perceptual experience, is itself a part of what is at stake in the contest between competing research programs. Victory in such a contest—victory *pro tem*, anyway—is a matter of a program's comparative success in

(1) providing a systematic means/regime/practice of experimentally *indexing* the candidate theoretical framework at issue, and
(2) providing systematically accurate *predictions/anticipations/explanations* of those same experimental indexings, as they are variously yielded in the context of relevant auxiliary indexings (i.e., 'initial conditions') in the cognitive context of the background theoretical framework as a whole.

The reader will perceive that I am here reexpressing Lakatos's and Kuhn's strictly linguaformal epistemological story in terms of the Sculpted Activation-Space or High-Dimensional Map account of theories on display throughout this book. The principal difference between the present account and these earlier accounts lies in

(1) the high-dimensional neuronal kinematics here assumed for human (and animal) cognitive activity in general,
(2) the vector-processing dynamics here assumed for perceptual activity in particular, and
(3) the recurrent modulation (of that vector-processing dynamics) here assumed for the exploration of diverse possible interpretations of our preconceptual sensory inputs, which inputs are simply the gross patterns of activation levels across a proprietary population of sensory neurons—visual, auditory, tactile, and so forth.

We are here deliberately looking *past* the blindingly familiar kinematics of sentences/propositions/beliefs, and the profoundly superficial dynamics of logical inference. We are looking past a kinematics and dynamics that jointly constitute the basic framework of traditional epistemology.

What, then, do we see? We see a framework for cognitive activity that is basically uniform across human and nonhuman brains alike, a framework that is uniform again across diverse human cultures, historical periods, and contending research programs. We see a framework for cognitive activity that has no initial commitments as to how one's perceptual activities should be conceptualized in the first place. And we see a framework in which any creature's anticipations of future experience are produced by a process far more intricate and sophisticated than can be captured by the strict deduction of observation sentences from any general propositional assumptions. Those singular anticipations are produced, instead, by a process of prototype activation that yields a dramatically *graded* field of more-or-less typical expectations, as opposed to the rigid and well-defined logical entailments of a discursive propositional hypothesis. Those graded anticipations—which are produced via a matrix of synaptic connections typically numbering in the hundreds of millions—reflect an acquired familiarity with, and an inarticulate expertise concerning, the domain at issue, an expertise that is typically far too intricate to be captured explicitly, or even relevantly, in a set of sentences.

We see a long-term process in which the accuracy or integrity of a cognitive map, as a map of a given objective feature-domain, is evaluated by its overall success in repeatedly generating local expectations that agree with the subsequent spontaneous or automatic indexings of its high-dimensional categories, indexings induced by whatever sensory equipment or measuring instruments the creature possesses. As we saw in the preceding chapter, one and the same map can often be indexed by two or more distinct sensory modalities (as when one can both see and feel the shape of an object, both see and hear the motion of an object, etc.). Individual indexings themselves are thus subject to scrutiny and evaluation, and they in turn provide an ongoing means of evaluating the anticipatory accuracy of the background framework as a whole. One can also deploy distinct but *partially overlapping* maps, indexed by distinct sensory instruments, to test the overall coherence of one's cognitive take on the world.

We see a framework for cognitive activity that sustains a *correspondence* theory of our actual representational success, although the relevant representations are not remotely linguistic in character. And we see a framework for cognitive activity that sustains a *coherence* theory of how our representations are evaluated, by *us*, for their representational success or failure. Once again, however, the relevant forms of coherence are not linguistic in nature. Rather, they concern the mutually conformal character of diverse but overlapping conceptual maps.

Such a naturalistic and non-linguaformal take on the processes of world representation must confront the inevitable objection that epistemology is originally and properly a *normative* discipline, concerned with the principles of *rational* thought, with how we *ought* to evaluate our theories. Since we cannot derive an "ought" from an "is," continues the objection, any descriptive account of the *de facto* operations of a brain must be strictly irrelevant to the question of how our representational states can be *justified*, and to the question of how a rational brain *ought* to conduct its cognitive affairs. If these evidently normative matters are our primary concern, then descriptive brain theory is strictly speaking a waste of our time.

There is something to this objection, no doubt, but much less than is commonly supposed. An immediate riposte points out that our normative convictions in *any* domain always have systematic factual presuppositions about the nature of that domain. They may be deep in the background, but they are always there. And if those factual presuppositions happen to be superficial, confused, or just plain false, then the normative convictions that presuppose them will have to be reevaluated, modified, or perhaps rejected entirely.

In the present case, this possibility is not so distant as one might have assumed. For we have already seen compelling evidence that the basic unit of world-representation for brains in general, including human brains, is not the belief, or the proposition, or virtuous systems thereof, despite centuries of normative discussions naïvely based on this parochial assumption. Rather, the basic units of representation are the activation vector (for fleeting representations of the here-and-now), the sculpted activation space (for enduring representations of the world's timeless structure), and the tuned matrix of synaptic connections (for shaping those spaces in the first place, and for transforming one high-dimensional activation vector into another). If we hope to set standards for, and to evaluate, the 'virtue' or 'rationality' of such cognitive activity as is conducted within this novel kinematical and dynamical framework (i.e., most of the cognitive activity on the planet), we are going to have to rethink our normative assumptions from scratch.

A second riposte points out that a deeper descriptive appreciation of how the cognitive machinery of a normal or typical brain *actually* functions, so as to represent the world, is likely to give us a much deeper insight into the manifold ways in which it can occasionally *fail* to function to our representational advantage, and a deeper insight into what *optimal* functioning might amount to. I here appeal to an analogy I have appealed to before,[6] namely, our understanding of the nature of living things, and our normative understanding of what a *healthy* body, metabolism, immune system, developmental process, or diet amounts to. Even into the early modern period, we had essentially no understanding of biological reality—certainly none at the microlevel—although we did have an abiding interest in the avowedly normative matter of what Health amounted to, and in how best to achieve or maintain it.

Imagine now a possible eighteenth century complaint, raised just as microbiology and biochemistry were getting started, that such descriptive scientific undertakings were strictly speaking a waste of our time, at least where normative matters such as Health are concerned, a complaint based on the 'principle' that "you can't derive an *ought* from an *is*." I take it the reader will agree that such a complaint is, or would have been, profoundly, even tragically, benighted. Our subsequent appreciation of the various viral and bacteriological origins of the pantheon of diseases that plague us, of the operations of the immune system, and of the endless sorts of degenerative conditions that undermine our normal metabolic functions, gave us an unprecedented insight into the underlying nature of Health and its many manipulable dimensions. Our normative wisdom increased a thousand-fold, and not just concerning means-to-ends, but concerning the identity and nature of the 'ultimate' ends themselves.

The present complaint concerning the study of Rationality, I suggest, is a clear parallel. In advance of a detailed understanding of exactly what brains are doing, and of how they do it, we are surpassingly unlikely to have any clear or authoritative appreciation of what doing it *best* amounts to. The nature of Rationality, in sum, is something we humans have only just begun to penetrate, and the cognitive neurosciences are sure to play a central role in advancing our normative as well as our descriptive understanding, just as in the prior case of Health.

3   On the Evaluation of Conceptual Frameworks: A Third Pass (Intertheoretic Reductions)

One-shot indexings of sundry conceptual maps, or rather, large numbers of such indexings, are instrumental in the evaluation of our maps for representational accuracy, as we have seen. Such indexings are always corrigible, of course, and always deploy some map-or-other themselves, so

6. In the closing paragraph of the last chapter of my *Neurophilosophy at Work* (Churchland 2007a).

there can be no privileged set of authoritative indexings fit to serve as the unchanging foundation for all of our map evaluations. But with a variety of maps and a variety of sensory systems with which to index them, we manage to use each map, and each sensory system, in such a fashion as to keep all of the *other* maps and sensory systems honest. This often means modifying those maps, or our habits of indexing them, so as to repair a discovered incoherence.

On some occasions, however, sensory indexings play little or no direct role in our evaluations of our conceptual maps. Sometimes we can put aside the ephemeral "you-are-here" indications delivered by our sensory systems, and contemplate the intricate internal structures of two distinct conceptual maps considered as static wholes. Here the issue is whether the acquired structure of one of our maps mirrors in some way (that is, whether it is homomorphic with) some *sub*structure of the second map under consideration. Is the first map, perhaps, simply a more familiar and more parochial *version* of a smallish *part* of the larger and more encompassing second map? Is the first and smaller map a version that superimposes nicely on that smallish part of the larger map in such a fashion that it can naturally be construed as *portraying the same domain* as the relevant portion of the larger map, though perhaps in greater, lesser, or somehow different detail? If so, then we may have discovered that the objective feature-domain represented by the first map is in fact a specific *part* of the larger objective feature-domain represented by the second map. We may have discovered that one entire class of phenomena is in fact identical with a special subset of another, more encompassing class of phenomena.

Our scientific history is replete with such discoveries, or presumptive discoveries. Clerk Maxwell's mathematical portrayal of the interaction between electric and magnetic fields, so as to produce electromagnetic (EM) waves, yielded a portrayal of their propagative behavior that mirrored almost exactly all of the known properties of *visible light*, including its highly unusual velocity in a vacuum, its refractive, reflective, polarizing, and interferometric properties, and its energy-carrying character as well. What we have here is the systematic subsumption of one highly specific conceptual map by another, much more broadly encompassing, concep-tual map. What we had long been calling "light," it began to seem, is just one tiny part of the large spectrum of possible *electromagnetic waves*—a spectrum that reaches from radio waves, microwaves, and infrared light on the longer side of the narrow band of wavelengths visible to humans, down to ultraviolet light, X-rays, and gamma rays on the shorter side. Our idiosyncratic human eyes give us a modest access to a tiny portion of this vast territory, and this allowed our earlier conceptions of light to get off the ground and to mature, eventually, to the stage of geometrical optics and to the rudimentary wave theories of Huygens and Young. These latter are all nontrivial maps, to be sure, for the physics of *visible* light is already an intricate and engaging thing in its own right; but they pale in comparison to the much broader portrayal of reality embodied in the conceptual map of electromagnetic theory as a whole.

This famous case of interconceptual subsumption is an example of what twentieth-century philosophers of science came to call "intertheoretic reduction" and rightly regarded as one of the most revealing sorts of theoretical insight that science can hope to achieve. Such 'reductions' are also among the most beautiful of human achievements, for it is quite arresting to suddenly perceive an already alluring (if dimly lit) reality revealed, when the stage-lights finally go up, as but the partial outline of an even more intricate and glorious background reality. Such subsumptions are more engaging than even this metaphor can convey, for it is entire families of *general* features or abstract *universals* that are thus revealed as but a small part of an even larger and more intricate family of general features. Intertheoretic reductions, as even the Logical Positivists had it, typically explain not individual *facts*, but rather entire *laws of nature*, in terms of even more *general* laws of nature. On our view, they explain the *timeless relations between universals*, above and beyond the ephemeral relations between particulars.

• "So *that's* why the characteristic refractive indexes of transparent substances such as water, glass, and alcohol are all different," marvels Maxwell. "The velocity of EM waves is different within, but characteristic of, each kind of medium!"

• "So *that's* why lunar, planetary, and cometary orbits are all conic sections," exults Newton. "Moons, planets, and comets all display a combination of an inertial motion and a centripetal acceleration under a force that varies as $1/R^2$!"

• "So *that's* why the pressure of an ideal gas under compression shrinks to zero as its temperature approaches $-273°$ C," exclaims Boltzmann. "At that limiting temperature its massive particles have ceased their ballistic motions entirely and thus no longer exert any counterpressure by continually bouncing off the walls of the gas's container!"

• "So *that's* why the periodic table of the elements has the peculiar structure of chemically similar 'element-families' that it does," celebrates Bohr.

"Those families reflect the repeating orbital configurations of the electrons in the outermost electron shell of the successive elements within each family, which configurations are precisely what dictate any element's chemical behavior!"

And so on, for a great many other entirely real historical examples.

Such examples often, even usually, fall on deaf ears, or better, arrive to blind eyes, because few people are sufficiently familiar with both the target map being subsumed, and the grander map that does the subsuming, to appreciate the fell-swoop explanatory insight that the latter typically provides. If one pursues a career as a scientist in some field, one likely learns a small handful of such local-map-by-grand-map subsumptions specific to the history of that field, but even scientists seldom learn too very much beyond their own field of specific expertise. Alternatively, if one pursues a career in the history of science generally, one likely learns to appreciate a rather larger handful of such conceptual subsumptions: those of maximal fame and subsequent intellectual influence. But even here the historian's grip on any such case is usually much less firm than that of a working scientist whose particular field contains that subsumption as one of its standout intellectual developments. And essentially no one alive today has an exhaustive grip on each and every one of the interconceptual subsumptions displayed in the entire history of human scientific activity. That would require that one be both a practicing expert within, and a devoted historian of, every scientific discipline in the academic and professional pantheon. This is too much to ask, and the demand becomes ever-more unrealistic as the sciences continue to progress.

Inevitably then, any real individual's empirical grasp of the cognitive phenomenon at issue—interconceptual subsumption—is almost certainly limited to only a smallish subset of the disparate and disconnected 'display galleries' that collectively constitute the museum of our overall scientific history. And that individual's view of the general phenomenon will inevitably be further pinched by the local peculiarities of the idiosyncratic instances with which he or she happens to be familiar. The responsible *evaluation* of any general philosophical account of interconceptual subsumption—an account adequate to the many historical instances thereof—will thus require the expertise of many people and a consensus among diverse judges.

Simply thinking up a *possible* account of the phenomenon, by contrast, is quite easy. Indeed, it is all too easy. Consider a brief survey of the glib alternatives already out there.

1. *The Simple Deductive Account* A more general framework, G, success-fully reduces a distinct target framework, T, if and only if the laws and principles of G *formally entail* the laws and principles of T.

Difficulties: The typically diverse lexicons of G and T will preclude any formal entailments of the kind required, for purely formal reasons. Those lexicons need somehow to be connected to one another before any deduc-tions can take place.

2. *The Modified Deductive Account* A more general framework, G, success-fully reduces a distinct target framework, T, if and only if the laws and principles of G, *in conjunction with certain 'bridge laws' or 'correspondence rules'* that connect the lexicons of G and T, formally entail the laws and principles of T.

Difficulties: The target framework, T, is typically false, at least in its details, which is inconsistent with the joint assumption of truth for G plus the correspondence rules, and the deducibility of T therefrom.

3. *The Restricted-Homomorphism Account* A more general framework, G, successfully reduces a distinct target framework, T, if and only if the laws and principles of G, *in conjunction with certain limiting and or counterfactual assumptions*, formally entails a restricted set of laws and principles, *Image(T)*, still expressed in the lexicon of G, an 'image' that is closely *homomorphic* with the target set of principles in T. Correspondence rules thus play no part in the deduction proper. They need be seen as nothing more than ordered pairs of terms, which collectively indicate the focal points of the *T-to-Image(T)* homomorphism. However, if the homomorphism achieved is sufficiently close, those correspondence rules can often reappear as material-mode cross-theoretic identity-statements motivated and justified by that global homomorphism. (Full disclosure: I have defended this account myself, in Churchland 1979.)

Difficulties: This third account has many virtues, but like the first two it remains wedded to what has been called *The Syntactic View* of theories, the view that a theory is a set of sentences or propositions. That anthropomor-phic, linguacentric view leaves out of the account all cognitive creatures other than language-using humans, and it may be fundamentally misdi-rected even there. (See below.)

4. *The Model-Theoretic Account* A theory is not a set of sentences. It is, rather, a set-theoretic structure, specifically, a family of *models* in the sense deployed by Tarski's account of truth. (To connect with the Syntactic View, a theory is here being construed as that family of models each of which

would make the corresponding set of sentences come out true on that interpretation. It is a set of models that share a common abstract structure. That shared abstract structure is what is important for theory-identity, not the idiosyncratic language in which the theory might happen to be expressed.) Accordingly, and to get (finally) to the point, a general theory G reduces a target theory T if and only if each one of the models for T is a *substructure* of some model for G.

Difficulties: This account is essentially an 'extensionalist' version of the Modified Deductive Account in (2) above, and it will need to be qualified or weakened somehow (as was account 3) above, to account for the frequent failure of target theories T to be strictly true, and/or to be perfectly isomorphic with the relevant reducing aspects of G. That repair, let us assume, can be achieved, and this account may then be counted at least a worthy alternative to account (3), worthy for many of the same reasons. Still, the account, as it stands presently, is motivated by no evident theory of the nature of human and animal cognition, of how the mind or brain *represents* anything at all, let alone how it represents intricate set-theoretic structures. Account (3) was motivated by what has turned out to be a *suspect* (i.e., a sentential) account of the elements of cognition, but at least it had some motivation in this regard. The Model-Theoretic Account has none, and does not bother to try.

5. *The 'Logical Supervenience' Account* A general framework, G, reduces a distinct target framework, T, if and only if the facts-as-expressed-in-T *logically supervene* on the facts-as-expressed-in-G.

This needs some elaboration. The idea at work here is that the terms in G, and in T, are typically *defined* by the peculiar causal/functional profile assigned, to the referents of the relevant terms, by the laws and principles of the background theory in which they appear. Accordingly, if the causal/functional profile for a given term in T is exactly the same as (or is a proper part of) the causal/functional profile for some term in G, then the local or factual applicability of that G-term in any given situation *logically guarantees* (i.e., *semantically* or *analytically* guarantees) the applicability of the corresponding T-term. For the defining features of the latter are (said to be) included in the defining features of the former. And if *all* of the terms in T find exact profile-analogues among the profiles definitive for the terms in G, then the truth of G as a whole 'logically guarantees' (once again, *analytically*, not formally, guarantees) the truth of T as a whole. Hence the claim that the facts-as-expressed-in-T *logically supervene* on the facts-as-expressed-in-G.

Difficulties: The idea that the existence of parallel causal/functional profiles is crucial to achieving reduction is a good one, as is the idea that the meaning of theoretical terms is somehow rooted in those profiles. But the idea that those profiles literally *define* the relevant terms is an idea that would make all of the relevant laws *analytically* true, a status starkly incompatible with their role as empirical generalizations capable of supporting nontrivial predictions and explanations. And the idea that, for example, something's meeting the causal/functional profile characteristic of EM waves (with wavelength λ between 0.4 and 0.7 μm) *analytically guarantees* that it is a case of visible light, is twice insupportable. Even if there were analytic truths, the statement "Visible light is identical with EM waves (with λ between .4 and .7 μm)" would certainly not be among them. It is an ostentatious paradigm of a synthetic empirical discovery.

Moreover, the notions of 'analytic truth' and 'analytic entailment' are substantially empty and epistemologically misleading in any case, as the profession has been (slowly) learning ever since Quine's (1951) landmark paper, "Two Dogmas of Empiricism," over half a century ago. As well, this account fails to deal with the stubborn fact, noted twice above already, that even successfully reduced theories are typically *false*, at least in minor and sometimes in major detail, a fact incompatible with the claim of analyticity for the principles of the reduced theory *T*, and also with the claim of their 'analytic entailment' by the collective principles of the more general reducing theory *G*. In all, this account takes a couple of fairly good opening ideas and then wraps them up in a destructive and self-defeating package of philosophical anachronisms. Finally, and perhaps worst of all, this account recommits us to the view that a theory is a set of sentences.

If we are to get an adequate grip on the real nature of intertheoretic reduction, we need to put all of these pre-Cambrian ideas—from (1) through (5)—behind us, and pursue an account that is grounded in the brain-based cognitive kinematics and dynamics under exploration in this book. To that end, let us now consider just such an account.

6. *The Map-Subsumption Account* A more general framework, *G*, successfully reduces a distinct target framework, *T*, if and only if the conceptual map *G*, or some part of it, *subsumes* the conceptual map *T*, at least roughly.

More specifically, a more general framework, *G*, successfully reduces a distinct target framework, *T*, if and only if

(a) the high-dimensional configuration of prototype-positions and prototype-trajectories within the sculpted neuronal-activation space that constitutes *T* (a conceptual map of some abstract feature-domain) is

(b) roughly *homomorphic with*

(c) some substructure or lower-dimensional projection of the high-dimensional configuration of prototype-positions and prototype-trajectories within the sculpted neuronal activation-space that constitutes G (a conceptual map of some more extensive abstract feature-domain).

Moreover, if this homomorphism or 'overlap' between the two maps is sufficiently close, we may be justified in concluding that any pair of corresponding map-*elements* represent the very same objective feature, and indeed, that the entire feature-domain represented by T is just a proper part of the larger feature-domain represented by G.

Difficulties: None. Or, more honestly, at least none of those difficulties that plagued the accounts of reduction earlier on our list. Most especially, this account avoids the linguacentricity of the several syntactic/deductive accounts. And yet, it remains strongly motivated by an independently compelling theory of what human and animal cognition *consists in*, namely, the construction and deployment of high-dimensional neuronal activation-spaces, spaces that embody sophisticated maps of various objective feature-domains, spaces that represent the world in an entirely prelinguistic or sublinguistic way. Moreover, this account explicitly allows for sundry representational *defects* in the target framework T, and for strong limitations on the degree or character of the homomorphism obtaining between T and G. Accordingly, a good theory, on this account, can therefore provide an illuminating reduction of a partially faulty theory. This is a virtue, because most historical reductions follow this very pattern.

Despite its focus on sculpted activation-spaces, this map-subsumption account is in no way intended to demean or deny the importance and cognitive relevance of the explicit formal and mathematical deductions that typically appear in the scientific literature and in our advanced science textbooks, deductions designed to explain and illustrate the reduction of some target theory T to some more general theory G. For such deductive illustrations do indeed serve to *locate* the feature-domain of T within the larger feature-domain of G, and they serve to *reconstruct* at least the major structural/relational features of that target domain in terms of the more basic structural/relational features of the reducing domain. A deductive-nomological argument can therefore evoke, in a forceful and relevant way, precisely the sort of systematic reconception of the target domain that any intertheoretic reduction invites us to embrace. The critical relation in this reduction, however, is not a deductive relation between sentences, but a conformal relation between distinct conceptual frameworks, that is, a conformal relation between distinct high-dimensional structures or maps within a person's neuronal activation-spaces.

That conformal relation is inevitably much richer, in its many neuronal dimensions, than is the deductive relation displayed in the science textbooks, just as the two theories or conceptual frameworks involved, G and T, are themselves much richer than are the sets of textbook-sentences typically deployed to try to express each of them. The cognitive skills of a scientific adept—just like the skills of a professional golfer, or a professional musician, or a chess champion—far exceed what can be captured or expressed in a set of rules or explicit sentences. And for the same reason: the true basis of such skills lies in the acquired structure of a high-dimensional neuronal activation space, a space whose internal structure is typically far too complex to receive any but the most rudimentary or low-dimensional articulation by the skill's possessor.

And yet, articulate them we do, if only poorly and partially. For we need to engage in collective cognition with other adepts, and we need to teach at least the entry-level rudiments of our sophisticated conceptual skills to the next generation of thinkers, and to encourage their (inevitably slow) development into comparable adepts. This process will be examined in the next chapter.

The virtues here extolled, of the map-subsumption account of intertheoretic reduction, include an especially vivid illustration of why a popular take on the general nature of "scientific reduction" is wildly mistaken and desperately confused. The term "reduction" was originally deployed by philosophers of science, in analyzing the familiar historical examples thereof, because in so many of the relevant cases, the collective equations of the more general theory G—when conjoined with or modified by the counterfactual assumptions, limiting conditions, and 'bridge laws' relevant to the case at hand—literally *reduced to* (in the strict *algebraic* sense of substituting and collecting terms, deleting canceling factors, and so forth) the assembled equations of the narrower theory T.

This is a perfectly appropriate, if specialized, use of the term "reduction," of course, but to most people that term strongly connotes the operation or activity of "making small," or "shrinking down," a very different matter. To this unfortunate connotation, we must further concede the salient fact that many prominent intertheoretic reductions (though far from all of them) appeal to the sundry denizens of the *micro*world—that is, to molecules, atoms, and subatomic particles. This appeal tends to compound the earlier confusion and to fan the fears of belittlement and over-simplification. Accordingly, it is no surprise that most people in academia, if they have any conception of intertheoretic reduction at all, think of it as an operation that sucks the life blood out of its target phenomena. They think of it as an operation that leaves those phenomena shrunken, demeaned, and conceptually abused.

This construal is tragic, because the truth, of course, is precisely the opposite. In a successful reduction, the target phenomena are revealed to be especially interesting *instances* of some much more *general* background phenomena, phenomena that reach far beyond the comparatively narrow perspective of the target framework *T*. The causal and relational profiles of the target phenomena get displayed, by the reduction, as local or humanly relevant manifestations of the overarching causal and relational profiles that characterize a grander reality reaching far beyond the initially evident reality explicitly comprehended by the target framework *T*. The initial charm of those already appreciated local profiles, which may indeed be considerable, is thus dramatically *enhanced* by being made an integral part of an even greater ontological and behavioral glory. If anything, those phenomena are thereby made 'larger,' not 'smaller.' But this cognitive expansion is typically, and understandably, *invisible* to those who do not command the new and more comprehensive framework.

Examples abound. Newtonian mechanics—the three laws of motion plus the law of gravity—uniquely embraced the Copernican–Keplerian account of our local *planetary* motions, as being correct, from among the competing trio that included the very different Ptolemaic and Tychonic accounts. The broad dynamical story that it provided decisively settled that three-way competition: only the Copernican–Keplerian account slipped smoothly into the larger Newtonian embrace. The planetary motions at issue were thereby revealed as so many local manifestations of patterns that hold throughout the entire physical universe, patterns that also embrace Jupiter's moons, binary stars, comets, and other solar systems.

The heavens aside, that very same Newtonian mechanics, plus the assumption that matter was made of molecules, also fixed the kinetic or molecular-motion account of heat and temperature as the uniquely correct account of *thermal* phenomena in general. The behavior of gases, the work-ings of steam engines, and the laws of classical thermodynamics generally (that is, the basic underpinnings of modern civilization's Industrial Revolu-tion) were thus revealed as further manifestations of the very same laws that governed the motions of the heavenly bodies. In all, Newton's theo-retical vision gave us an unprecedentedly fertile and unified conception of physical phenomena from the scale of moving molecules up through the scale of flying cannon-balls and revolving planets and on to the scale of entire galaxies of stars revolving by the millions in spiral concert. There is nothing in any of this that tends to demean or to demote the various phenomena thus enfolded in the Newtonian embrace. Just the reverse.

A comparable expansion of our perspective is displayed in the case of the reduction of light to electromagnetic waves, as was briefly discussed several pages ago. And aside from the assimilation of light itself to the full range of electromagnetic waves criss-crossing the universe, all at the same velocity but at widely different wavelengths and energies, there is the still broader context of *nontraveling* electric and magnetic fields, with their endlessly various manifestations in the behavior of compass needles, electric currents, electric motors, telegraphs, telephones, radio circuits, television screens, computer chips, spark plugs, lightning bolts, electric power stations, and the Aurora Borealis. Visible light is but one glistening facet of a high-dimensional jewel with many other facets as well, each one a distinct window onto the same underlying reality. If you loved light before—and who does not?—you will love it even more after all these related lessons have been learned.

This frankly revelatory pattern is the rule, not the exception, with successful interconceptual reductions. And further historical examples can be used to make the same point. Classical chemistry proved to be a real but still modest window onto the much larger domain collectively addressed by modern atomic theory, structural chemistry, and modern thermodynamics, a larger theoretical domain that also embraces the diverse phenomena of biology—especially the phenomena of reproduction, metabolism, and evolution. Here we are contemplating, in a unified fashion, the structural articulation of matter at the hands of energy, at all scales, both spatial and temporal.

But enough heartwarming or awe-inspiring examples. Our principal concern in this section is with the *epistemological* significance of so-called 'intertheoretic reductions,' not with their aesthetic impact on those with eyes to see them. And that epistemological significance resides foremost in the fact that, if we have presumptively *overlapping* pairs of maps, then we are in a position to compare their respective depictions or portrayals of their shared domain, and to evaluate them for their accuracy, their comparative detail, and their possibly diverse representational concerns. We have been doing this for centuries, without fully appreciating just what we were doing.

4   Scientific Realism and the Underdetermination of Theory by Evidence

The account of cognition here under examination has earlier been claimed, by me, to be a novel incarnation of what philosophers of science have long thought of as Scientific Realism. Specifically, and on the present view, a theory is a conceptual framework or high-dimensional cognitive map that purports to be, and sometimes succeeds in being, an *accurate* or *faithful* map of some domain of objective features and the enduring relations that hold between them. It remains for us to examine one of the traditional difficulties for Scientific Realism, namely, the presumed underdetermination of theory by empirical evidence. How does that familiar drama play out on the novel epistemological stage here proposed? In particular, how do some of the familiar *anti*realist arguments appear when restated within the neurocognitive kinematics and dynamics now before us?

Perhaps the first issue here is the status of the familiar 'pessimistic metainduction' concerning the historical achievements of science and the current prospects for truth in our contemporary theories. We may state the argument as follows.

All past scientific theories, widely embraced on the evidence available at the time, have turned out to be either entirely or substantially false. Therefore,

All of our current scientific theories, even though widely embraced on current evidence, are likely to be either entirely or substantially false. Indeed, and by the same token, any future theory embraced on evidence then-available is also likely to be either entirely or substantially false. Therefore,

We ought not conceive of science as an enterprise whose proper aim is truth. It should be conceived as an enterprise with some more modest aim.

Here, a variety of more modest aims are typically proposed. For example, the Pragmatist recommends mere Practical Effectiveness as a more appropriate aim for science. The Constructive Empiricist recommends mere Empirical Adequacy as a more appropriate aim. I have argued against this Pragmatist proposal earlier in this book (pp. 128–134), and I argued at length against Constructive Empiricism some twenty-five years ago (see Churchland 1985), so I will here spare the reader any detailed critique of these existing alternatives. For I mean to propose and defend a new and competing alternative as the essential business and the proper aim of science: namely, high-dimensional Cartography, in the first case, and the construction of ever-more accurate and comprehensive cognitive maps of the enduring structure of reality, in the second.

The proper aim of science, on this account, is arguably even more ambitious than is the aim embraced by the classical scientific realist, namely, simple linguistic truth. This is because the medium of continuous-valued, interacting neuronal maps of arbitrarily high dimensionality is arguably a richer medium of potential representation and computation than is the discontinuous, concatenative medium of human language. At first, this might seem to make a problematic situation even worse, rather than better, for the optimistic scientific realist. For the bar has now been raised, it would seem, even higher than the goal of linguistic truth. A reformulated metainduction, grounded in the sundry representational failures of our past theories, might thus leave us even *more* pessimistic about the prospects for representational success in our current best theories.

But in fact, the reverse is true, because construing a conceptual framework as a kind of map allows us an importantly *graded* mode of evaluation that is mostly missing in the bivalent case of sentences and their truth or falsity. Aristotle's Law of the Excluded Middle—for any *P*, either *P* is true or *P* is false, but not both—lies heavy upon our evaluation of any sentence, or any set of them. Informally, we occasionally do try to speak of 'degrees of truth' or of 'partial truth,' especially in cognitively awkward circumstances. But such talk is unsystematic, hard to sustain, and strictly at odds with the ruthless bivalence imposed by Aristotle's Law.

If, however, we construe scientific understanding as the possession of sundry *maps* of the enduring categories, symmetries, and invariants displayed by the objective universe, then we are free to evaluate the successes or failures of those maps in a large variety of distinct *dimensions*, and to evaluate those maps as varying in continuous *degrees* of accuracy and inaccuracy. We can then acknowledge, as surely we must, the manifold representational failures displayed by the many 'temporarily triumphant' theories that populate our long scientific history, while still maintaining (a) that each of them boasted at least some representational success in at least some dimensions of evaluation, however narrow, and (b) that successive theories in the same or overlapping domains often show a still greater representational success across a yet wider range of dimensions than did their rightly superceded predecessors. In fact, this is precisely what the successful map subsumptions (a.k.a. intertheoretic reductions) discussed in the previous section provide us. They typically give us a (usually only partial) *vindication* of the ontology and the world-portrayal of the reduced theory, plus a systematic *correction* of the elements and/or the structural relations of that earlier world-portrayal. A poor map gets subsumed by a larger and better map, whose internal structures provide both an explanation of why the older map has the peculiar structural features it does, plus a template for their systematic correction.

When we see the unfolding history of our theoretical commitments through these cartographical lenses, its inductive lessons are far more optimistic than the gloomy conclusion drawn by the classical metainduction that focused on linguistic truth. The bivalent straightjacket of classical linguistic logic forced us to acknowledge the strict falsity of even our best theories at every stage of human history. We are thus forced to infer the probable strict falsity of even our best *current* theories as well. I accept that inference and I embrace its conclusion. I think we must. But that same epistemological modesty, regarding truth, is entirely consistent with a different and much more optimistic lesson that is *also* afforded by our scientific history. A detailed look at our history shows, no less robustly, that our conceptual maps of enduring reality have *improved* dramatically in both their breadth and in their accuracy. Their serial imperfections and failures conceded, they have given us progressively *better* portrayals of the reality that embeds us, in all of the evaluatory dimensions germane to any map.

I therefore propose, as a complementary adjunct to the familiar pessimistic metainduction concerning the truth of even our best theories, the following deliberately *optimistic* metainduction concerning the representational virtues of our currently favored neuronal *maps*.

All of our past neuronal maps, when widely embraced on the strength of their comparative performance at the time, subsequently turned out to be at least partly accurate portrayals of at least some dimensions of reality, *even as judged from the stern perspective of the superior neuronal maps that actually displaced them.*
Therefore,
All of our current neuronal maps, widely embraced on the strength of their comparative performance at the present time, are also likely to be at least partly accurate portrayals of at least some dimensions of reality, even as judged from the unseen perspective of the neuronal maps that will eventually displace them. Indeed, and by the same token, any future neuronal maps, if widely embraced on the strength of their comparative performance at that time, are also likely to be at least partly accurate portrayals of at least some dimensions of reality, even as judged from the perspective of whatever neuronal maps will eventually displace them as well.

Therefore,

We can indeed portray the enterprise of science as one of seeking ever-more comprehensive and accurate conceptions/portrayals of objective reality, and as having achieved at least some success in this regard.

This argument is not, and does not pretend to be, a 'transcendental' argument for the success of science. It is a humble empirical argument, and it is entirely consistent with a robust suspicion that even our current best theories are false in ways that are currently beyond our imagination. It also stops short of the assumption, common among Scientific Realists, that the scientific enterprise is fated, in the long run, to *converge* upon a single theory, or single family of theories. It may well. But it need not. This is not because the convergence of complementary portrayals is unimportant. On the contrary, it is extremely important: it is perhaps the single best measure of representational progress we can deploy. Even so, it remains possible that there be *more than one* long-term focus or target of representational convergence. We shall return to this point in the next section.

The underdetermination of theory by our inevitably finite evidence has long been a presumptive problem for the aspirations of scientific realism. On the syntactic view of theories (a theory is a set of sentences), it is impossible to escape the fact that, for any finite set of 'data sentences,' there always exist indefinitely many distinct possible theories (i.e., sets of sentences) each of which stands in the relation of 'explaining' and/or 'predicting' those data sentences. And on the semantic view of theories (a theory is a family of models), there are always a great many distinct possible families of models, all of which share a common 'empirical substructure,' which substructure is the only aspect of those larger realities to which we humans are supposed to have any observational access. Thus, and on either account of theories, no amount of observational evidence can possibly serve to disambiguate the theoretical alternatives that are ever and always out there. Well and good. But let us ask, once again, how does this familiar bogey of underdetermination present itself within the quite different kinematical/dynamical framework under exploration in this book?

A theory, to reiterate, is a *sculpted activation-space* that purports to map the similarity-structure of the enduring features of some objective domain. But what is 'evidence,' within this picture? Or, for that matter, what is 'counterevidence'? One might have thought that the natural candidate for evidence here, whether positive or negative, is the individual activation vector, the singular map-indexing, the sensorily highlighted 'point' in the already-sculpted background activation-space. And indeed, this construal, perhaps, comes closest to philosophy's traditional candidate for the role of providing evidence, namely, the singular observation judgment.

But the fit is marginal. Activation vectors have no logical structure. They stand in no logical relations to the high-dimensional maps in which they are spatially located. And they are not, strictly speaking, either true or false. Moreover, their claim to *represent* anything in the world is entirely parasitic on, and contingent upon the success of, the antecedent representational virtues of whatever background map or *theory* is the locus in which those fleeting activations occur. That background, recall (from chapter 2, section 8), is where they get their semantic/representational content in the first place. They are therefore irredeemably theory-laden themselves, and thus unfit to play the role of 'theory-neutral' arbiters of theoretical truth and falsity. They are a hopelessly entangled *part* of the very conceptual practices we aspire to evaluate.

On the other hand, a singular sensory indexing does not *need* to be theory- *neutral* in order to have a major evaluative impact, either negative or positive, on one's current cognitive commitments. If the normal applications of one's conceptual framework repeatedly yield local sensory indexings ('observations') that, while comfortably expressed in the very framework at issue, are plainly *distinct from* the sensory indexings that one's background understanding of the current empirical situation had antecedently led one to *expect*, then something is clearly wrong. The overall cognitive system is in a state of at least a temporary disequilibrium. Either one's background map is itself inaccurate, or one's general capacity to index it correctly is out of whack, or one's assumptions about how to apply that background map to the local circumstances are in some way faulty. *Which* of these several possible failings is the culprit behind the dissonant sensory indexings at issue can and often will be a matter of considerable sleuthing on the part of the relevant cognitive agent. The Popperian impulse to see any such dissonant sensory indexing as a presumptive 'falsifying observation' is a romantic oversimplification, as we saw earlier. Even so, such dissonant indexings do command one's attention; they do herald a cognitive failing *somewhere* in the cognitive system; and they do motivate a reevaluation of one's existing cognitive practices and cognitive commitments. What they do *not* do is constitute a unique class of specifically *evidential* items against which all other elements in one's cognitive life are to be asymmetrically evaluated.

This is because they, too, can be evaluated for their integrity, and rejected if they are found wanting. And what will motivate such occasional rejections is the *evidence* provided by the accumulated authority of one's background conceptual framework, and the additional confidence that its local anticipations of impending experience are well-placed. In such cases, it is the sum total of one's *background* commitments that constitute the evidence relevant to evaluating the integrity of the presumptive 'observation' at issue, not the other way around.

This point is reinforced by the avowedly evidential role often played by entire *theories* in evaluating the credibility not just of observations, but of other theories as well. Think, for example, of the credibility of the Creationist's 'young Earth' hypothesis, and its correlative theory that all biological species originally appeared, as fixed and unchanging kinds, some 6,000 years ago. That covey of conjectural commitments, let us agree, is profoundly unlikely. But the primary evidence against it consists in the individual and collective authority of a number of other *theories*, namely, Darwin's theory of natural selection, deep historical geology (especially sedimentary geology), the biologists' reconstruction (from the fossil record) of species' appearance and extinction, the biochemists' reconstruction of terrestrial evolutionary history from both nucleic and mitochondrial DNA, planetary astronomy and its long developmental timelines, nuclear physics and its deep-time radioactive dating techniques, and our growing armory of biochemical techniques for the genetic modification of simple organisms and the in vitro exploration of their artificial evolution over time.

In this case, the 'evidence' against the Creationist hypothesis is far more telling than just the assembled observational evidence that tends to support these other theories. This high-level 'evidence' is the more telling because these other theories form a coherent and interlocking family of conceptual maps, maps that display an unprecedented level of success in allowing scientists to navigate their respective domains, maps that claim an even greater authority to represent our geological and biological history by virtue of their conforming *to each other* in such systematic and revealing ways. The principal evidence against Creationist biology is not just a few problematic observational facts, nor even a teeming number of such observational conundrums. It is a hulking giant of an alternative interpretation of the 'empirical data,' a hugely successful family of distinct but mutually conforming explanatory *theories*. Indeed, the 'observational data' that scientists report—such as "In this sedimentary layer we see a typical Trilobyte, from the late Ordovichian period, some 450 million years ago"—derive their authority, in large measure, from the earned authority of the background theory in whose concepts they are expressed.

What happens to play the role of 'evidence,' then, is a highly contextual matter, and general theories can and regularly do play that role no less than specific observations, depending on the circumstance. Indeed, the very notion of 'evidence' may have its proper and primary home in overtly *dialectical* situations, such as a courtroom or a scientific colloquium, where human interlocutors contend for and against some verbally articulated thesis. To force that notion onto the mostly sublinguistic process of Second-Level Learning may be to misrepresent both the kinematics and the dynamics of this important form of cognitive activity, even in language-using humans.

An alternative, nonlinguistic, and *prima facie* worthy view of Second-Level Learning lies already before us. A brain that has already reached rough maturity, in its history of *First*-Level Learning, is then in the position of trying to respond to its ongoing sensory inputs, with second-by-second sensory indexings of its background conceptual maps, in such a fashion as to maintain a *coherent* representational narrative of its unfolding adventures. Recall that the brain's most important maps are likely those for typical causal *processes*, and when the early stages of these causal prototypes are sensorily activated, as we saw, the brain automatically becomes the site of an unfolding family of *expectations* concerning the later stages of the relevant process. Those internally generated expectations will usually agree with the externally generated *sensory* indexings of the relevant external causal process, at least if the brain has been well trained. Indeed, the ideal situation is where the brain's expectations are *always* in conformity with its subsequent sensory indexings; that is, where the creature always anticipates its experience correctly.

But that mutual conformity, in fact, is mostly only partial. One's perceptions are regularly occluded, interrupted, or distracted. Accordingly, one's background prototypes and recurrent pathways are continually trying to make good on missing perceptual information or confounding noise, as we saw for the Predictor Network above. Moreover, the expectation-generating mechanisms, in animals, are chronically noisy in their operations, further compromising their conformity with perceptual inputs. These factors aside, the brain's background map of typical causal processes is itself imperfect, and so flawless conformity to its perceptual inputs is unlikely, even if those inputs were to be perfect and the background system were to be wholly free of noise. Finally, the objective causal processes encountered in sensory experiences are themselves endless variations on prototypical themes, for in the real world, no two causal situations are perfectly identical.

A brain, therefore, even if fairly well-trained in its global synaptic configuration, is fated to lead an interesting and ever-novel life in trying to maintain a smooth cognitive equilibrium in the face of the preceding complexities. Those complexities may be modest, for the most part, but they are never-ending. This invites a conception of the brain as a robustly homeostatic dynamical system, struggling to maintain its inevitable deviations from cognitive equilibrium within some acceptable limits, as it tries to navigate a reliably nourishing but endlessly noisy epistemological environment. The metaphor that leaps to mind is that of an ongoing *cognitive metabolism*, stable against minor variations in the nourishment it receives, and against minor abnormalities in the performance it displays. Indeed, this may be no mere metaphor. The dynamical profile displayed by the brain and nervous system is evidently another instance of the broader dynamical profile displayed by living things generally. Each of these systems is embedded in an environment of available energy and available information; each needs to maintain at least a minimum operational stability in that sometimes turbulent environment; and each exploits its operational stability so as to undergo substantial growth and development. We shall return to this theme a little later.

Minor disturbances to the brain's cognitive equilibrium are mostly unproblematic. As we saw, their causes are mostly humdrum and the brain's neuronal configuration is well-equipped to cope with them smoothly. But sometimes the disturbances are major, and recurring, and they disrupt one's practical affairs as well as one's inner reflections. Where they occur in the scientific realm, these recurring disturbances in cognitive activity have come to be called "anomalies," following T. S. Kuhn's penetrating analysis of their role in our sometimes revolutionary scientific history (see Kuhn 1962). Kuhn's idea was that these discovered failures—to anticipate, control, and explain some domain of phenomena hitherto thought to be unproblematic—often herald the need for a major shift in the way that these phenomena have been conceptualized. In neurocognitive terms, those anomalies may indicate that the high-dimensional map currently deployed in their habitual comprehension *is the wrong one*, and that we need to apply a wholly different high-dimensional map, drawn from elsewhere in our existing conceptual repertoire, in order to gain a new and better cognitive grip on the problematic domain at issue.

When two such competing maps contend, within the relevant scientific profession, for the right to interpret the problematic empirical record, Kuhn speaks of "Crisis Science." And when an older map, after a stressful period of conflict, finally gets displaced by some newer map, to serve as the default conceptual resource of that profession in that domain, he speaks of "Scientific Revolutions." Kuhn has occasionally been criticized for overdramatizing matters here, on grounds that episodes of reconceptualization take place at all scales, from broad-scale social revolutions to private epiphanies, and from the intellectually spectacular to the comparatively parochial and mundane. And so they do. But Kuhn's insights here are entirely real, according to the present account. His principal focus, of course, is at the social level, which we shall address in the next chapter. But in the early sections of this chapter we saw many examples of individual scientists undergoing a discontinuous shift in their interpretation of some familiar phenomenon. A comparable shift in the cognitive behavior of an entire community is bound to be somewhat slower, given the resistance that is almost always involved.

The historical episodes addressed by Kuhn have a further significance in the present context. As philosophers such as Feyerabend and Kuhn have become celebrated for insisting, all of the interesting theoretical decisions made by humans throughout our scientific history have been the result of contests between two or more *theories*, theories that were competing for the right to *interpret* the so-called 'empirical data,' rather than the result of solo attempts to make a single theory conform to some stable and unproblematic domain of sensory comprehension. And as we have seen from the earlier sections of this chapter, evaluating the representational integrity of our theories—that is, our cognitive maps—is always and ever a matter of playing them off, with and against each other, to unearth discrepancies and failures, and to solidify conformities and successes. This means that the traditional issues concerning the underdetermination of theories by evidence have been substantially misconceived. They need to be reformulated within the new framework on offer.

5   Underdetermination Reconceived

So how do the traditional issues reappear within this new framework? Rather differently, both in the nature of the problems posed, and in the surprisingly reassuring nature of the answers they invite. To begin, let us ask: if two (or a thousand, or a billion; it hardly matters) closely similar infant human brains are allowed to develop their cognitive resources in the *same* physical and social environment, will they converge upon the same family of high-dimensional maps of the environment's lasting structure, enduring universals, and typical causal processes? This is an experiment that has been performed countless times in human history, and is still being performed, even as we speak. And the robustly empirical answer is that, yes, indeed they do converge—very, very closely, if never quite perfectly.

Conceivably, this empirical fact is the reflection of the biological fact that our concepts are simply innate, and not the result of long development in response to individual experience. But this familiar suggestion, as we saw, runs afoul of the nine-orders-of-magnitude gap between the coding capacity of our genome (20,000 protein-making genes) and the number of synaptic connections (10^{14}) that require individual and highly variable weighting in order to produce a functional conceptual framework. No, the pluripotent brains of human *infants*, at least, are clearly converging—toward closely similar frameworks—in response to something beyond our genes.

Perhaps, then, it is our surrounding culture, and most centrally, our shared language, with its hierarchically organized vocabulary and its web of commonly shared assumptions. These can form a highly influential *template*, at least, around which any child's conceptual maps will be constrained to develop. This suggestion carries a good deal of weight, as we shall see in the following chapter. Our evolving conceptual culture does form a powerful attractor that pulls in each new generation of learning brains. But even here, and duly conceding the influence of culture, human brains are still responding to a shared perceptual environment, even if some of it is a social environment, and they still converge on a shared family of high-dimensional maps. It is this robust tendency to conceptual convergence that continues to stand out.

Indeed, it stands out even in the case of animals that have little culture and no language at all. Mammals and birds apparently arrive, at least within a species and within a common environment, to roughly the same conception of a three-dimensional universe with the same taxonomy of mostly unchanging kinds of physical objects enlivened by a special subset (animals) that display characteristic kinds of causal and motor behaviors. If we are to estimate the conceptual resources of these nonlinguistic animals by what is necessary to explain their behavior in its natural setting, their take on the world is much the same as, if somewhat simpler than, our own. No doubt this reflects the fact that they have much the same suite of sensory organs that we do, and the fact that the process of map-generation follows the same Hebbian process in them as it does in us. Recall the feedforward network that spontaneously developed Eleanor's girlish face as one of its 'preferred patterns,' or the recurrent network that spontaneously developed the flying bird as one of its 'preferred processes.' Hebbian learning is fiercely empirical its behavior. Place the same sensory systems in the same statistical environment, and Hebbian learning will steer the higher rungs in the processing ladder toward the same, or a closely similar, cognitive configuration every time. It happens in real creatures, and it happens also in our artificial network models (see, e.g., Cottrell and Laakso 2000).[10]

But all of these reassuring points concern *First-Level Learning*, you will rightly object, and our primary concern in the present chapter is with the process of *Second-Level Learning*, a rather more volatile and much less predictable process, since it involves the dynamical adventures of a strongly nonlinear dynamical system. Moreover, the comforting conclusions of the preceding paragraph depended on the assumption that the distinct brains whose conceptual convergence was at issue shared the same sensory organs and deployed them in the same environment. What happens if we lift our sights beyond these comparatively easy cases?

Several interesting things. First, creatures with sense organs very different from ours will of course develop conceptual maps, for sensory indexing, quite different from our own. They will be spontaneously and systematically accessing different aspects of the objective world, aspects to which we are blind, deaf, and so on. Since there are a great many different aspects of the world to which some creature or other might become sensorily attuned (think of being sensitive to X-rays, for example, or to magnetic fields), and since there are at least as many different sensory systems that might be used to achieve such attunement, we should positively expect the kind of conceptual diversity here contemplated.

But no interesting skeptical consequences flow inevitably from these assumptions. We humans are ourselves the locus of a diverse family of sense organs tuned to radically different aspects of the world. But we manage to construct a unified conception of reality even so. For our sensory modalities *overlap*, to some degree, in their several objective domains. Some aspects of reality can be seen, heard, felt, and smelled all at the same time (think of a campfire, for example), and thus our various conceptual maps can be used to confirm or contraindicate each other's spontaneous indexings, and thus steer us toward a stable representation of the local reality.

This benign situation changes in no fundamental way when we introduce *artificial* instruments of detection and measurement to our 'sensory armamentarium,' instruments such as voltmeters, ammeters, barometers, X-ray machines, radiometers, sonograms, magnetometers, gamma-ray telescopes, and infrared cameras. Indeed, it typically improves our situation dramatically, since these artificial devices for indexing novel abstract maps simply provide us with that many more independent but partially overlapping 'witnesses' to the unfolding reality around us. They positively *serve* the interests of conceptual convergence, rather than frustrate it, for the same reasons that the diversity of our native sense organs serves that aim.

On the other hand, this cheerful assessment ignores the dimension of purely *conceptual* diversity in the possible maps we can use to interpret even a *single* measuring instrument or sensory modality. It is precisely this dimension of diversity that is exploited in the process of Second-Level Learning, and I can illustrate the nature of the philosophical difficulty it poses with a true story about a conceptual/perceptual crisis recently confronted by the entire nation of Canada. In the mid-1970s, the Canadian parliament decided to get the jump on the other English-speaking nations of the world—most of which use the Imperial System of weights and measures—and switch *all* of Canada's industrial, economic, educational, and social activities to the plainly superior Metric System used by everyone else on the planet. A switchover date was set, and on that morning the people of Canada woke up to a new universe. Milk was sold by the liter, not the quart. Sausages by the kilogram, not the pound. Mileage signs on the nation's highways had all come down, replaced by unfamiliar distances in kilometers. Speed-limit signs everywhere displayed kilometers per hour. Gas was sold in liters, not gallons. New-car mileage was announced in liters per 100 kilometers (yes, that's right: not only are the units different, but their order has been inverted). Meters and centimeters replaced yards and inches in the schools and in the hardware stores. The local radios and newspapers were all required to announce the daily atmospheric temperatures in degrees Celsius instead of degrees Fahrenheit. Chaos and confusion reigned everywhere.

This last requirement, concerning temperature, was especially confusing. Not just because the Celsius degree is almost twice as large as a Fahrenheit degree, but because the two most natural signposts—the freezing and boiling points of water—are very differently placed on the two scales. Water freezes at 0° Celsius and boils at 100° Celsius (instead of 32° F and 212° F, in case you are unfamiliar with the Fahrenheit scale), which means that a pleasant summer afternoon will have a temperature of around 20° Celsius. But the expression "20°," to an American or (then) Canadian immediately conjures up thoughts of a bitterly cold *winter* day, with the temperature well below freezing. Similar confusions attended other signpost temperatures.

Frustrated paradigms aside, the Fahrenheit scale was then an automatic framework for the spontaneous and automatic *perception* of temperatures, for everyone in North America. Any normally enculturated citizen of Canada was able to judge the ambient atmospheric temperature, quite reliably, to within plus-or-minus two degrees Fahrenheit. Fahrenheit temperature was simply an observable feature of the universe. But nobody had yet acquired the skill of doing the same thing within the framework of the new Celsius scale. Indeed, the inertia of the old habits of sensory indexing positively stood in the way. For almost two years, no one in Canada was quite sure what the temperature was! (A serious matter for a country with such severe winters: I was living in Winnipeg at the time, the coldest major city on the planet.) And if you did happen to report the temperature in degrees Celsius, people would whisper, "What's that in *real* degrees?"

Withal, the old framework eventually faded away, and almost everyone learned to make spontaneous perceptual reports within the Celsius scale. Peace returned to the land. But not to me. While happy at the modest conceptual updating that had occurred, I remember feeling frustrated at the outset that our parliamentarians had foisted upon us the Celsius scale instead of the *Kelvin* scale of temperature. (The Kelvin scale has degrees equal in *magnitude* to degrees Celsius, but its zero point is *absolute zero*, a temperature at which molecules cease to move entirely. That is absolutely the lowest temperature possible, because the energy of molecular motion is ultimately what temperature *is*. That rock-bottom temperature is equal to –273° Celsius.) On this third scale, water freezes at 273 kelvins and boils at 373 kelvins.

My frustration was owed to the fact that all of the important laws of thermodynamics—starting with the classical gas law, $PV = \mu RT$, but hardly ending there—require that the number entered for T in such laws must be in kelvins, that is, in degrees above that unique and objective minimum: absolute zero. I reasoned that, if we humans could learn to make spontaneous and accurate perceptions of both Fahrenheit and Celsius temperatures, as obviously we can and did, we could just as easily learn to frame our spontaneous perceptual judgments in terms of Kelvin temperatures. We would then be in a position to simply substitute such perceived values directly into the various laws of thermodynamics, which would then give us a quick inferential access to all manner of interesting empirical facts about the local environment. (I'll spare you the physics lesson.) We would then be lifting our sensory indexings into the context of a rather richer conceptual map of the principal relations uniting the various universals that make up the thermodynamic domain, and both our conceptual insight into and our perceptual access to those universals would be thereby enhanced, as would our capacity for manipulating them.

This insight, if that is what it was, didn't last a day before I realized that, if we could learn to use the Kelvin scale as a spontaneous observational framework, we could just as easily learn to use the *Boltzmann* scale in that role, which still *more* penetrating scale characterizes directly the mean molecular kinetic energy of any gas, that is, the average kinetic energy of the individual molecules that make up that gas. I pulled down my old thermodynamics textbooks, dusty from years on the shelf, and quickly reconstructed the range of molecular kinetic energies to which we humans are reliably sensitive. We are perceptually reliable within a Fahrenheit range of roughly –20°F to +120°F, and within a Celsius range of roughly –30°C to +50°C. This corresponds to a molecular-kinetic-energy range of roughly 5.1×10^{-21} joules to 6.8×10^{-21} joules. As well, we should be able to spontaneously judge the values in this range to an accuracy of plus-or-minus 0.025×10^{-21} joules, which corresponds proportionally to our accuracy in the other two scales.

Who would have thought that such submicroscopic intricacies as these could literally be at one's sensory fingertips? I set about applying this scale—named after Ludwig Boltzmann, the great theorist of the kinetic theory of heat—in spontaneous observation, as the rest of Canada was still struggling to apply the Celsius scale. Within a few weeks of practice I could hold up a questing finger in a slightly overheated room (say, 78°F) and announce, "The kinetic energy of the average air molecule in this room is 6.3×10^{-21} joules; let's turn down the thermostat." A midwinter morning's outdoor temperature (say, –12°F) would elicit, "The kinetic energy of the average air molecule is only 5.2×10^{-21} joules . . . brrrr," And so on. At the time, I drove some of my coffee-quaffing colleagues in the University College Common Room to distraction, not just with these cheekily 'oracular' observations, but with my eager explanations of how *they too* could learn to make them. But for their current and very real struggles with the Celsius scale, those worthies might well have rescinded my Common Room membership entirely.

You may think the story ends here. But no. The kinetic energy of anything, you will recall, is just its mass times its velocity squared, divided by two, or $mv^2/2$. And this is the feature, of an average air molecule, that I had learned to detect spontaneously. But the typical atmospheric air molecule is a unit of diatomic nitrogen, or N_2, whose mass is about 4.7×10^{-27} kg. If we divide this mass into $mv^2/2$ (at room temperature, this equals 6.2×10^{-21} joules), and multiply by 2, we are left with v^2. Take the square root of this and we are left with v, the typical velocity of a single N_2 molecule. Accordingly, the *velocity* of a typical air molecule is as sensorily accessible as is its kinetic energy, at least within the human range of thermal sensitivity. For an atmosphere at room temperature, for example, that velocity is about 500 meters per second. And such judgments, like those for the other scales, are reliable to within plus-or-minus 2 percent. You may not be able to feel the molecules themselves, but you can certainly feel how fast or how slowly they are moving!

You may note that this value—500 m/s—is fairly close to the speed of sound at room temperature: 343 m/s. No surprise. For reasons that I will pass over, the speed of sound in any diatomic atmosphere is always about 70 percent of the atmosphere's average molecular velocity. This means that our 'thermal sensations' are equally good indexes of the speed of sound in the locally embedding atmosphere, for the speed of sound varies rigidly with atmosphere temperature. Within this new scale, we are sensitive to a range of about 313 m/s at our lower limit, up to about 355 m/s at our upper limit. Who would have thought that the local speed of sound was something that could be *felt*? But it can, and quite accurately, too. Of what earthly use might that be? Well, suppose you are a bat, for example. The echo-return time of a probing squeak, to which bats have accurate access, gives you the exact *distance* to an edible target moth, *if* you have running access to the local speed of sound.

To complete this story, it is worth looking back in human history for a final example of a conceptual map for exploiting the indexing potential of our native 'thermal sensations.' The ancient Greeks apparently regarded *hotness* and *coldness* as entirely distinct but mutually opposing properties. Bring a hot thing and cold thing into contact and those two sensible features will simply cancel one another out, leaving you with two things, each of which is neither hot nor cold. If we add this more primitive conception to our growing list, we may summarize this exploration as follows.

Possible Frameworks for Conceptualizing 'Sensible Thermal Reality':
1. The ancient Greek framework
2. The Fahrenheit framework
3. The Celsius framework
4. The Kelvin framework
5. The Boltzmann framework (molecular KE)
6. The molecular-velocity framework
7. The speed-of-sound framework

One need not stop here: fantastical examples can be added to the list. But these seven entirely real possibilities will serve to make the relevant point, namely, that the mere range or profile of one's sensations within any sensory modality does *not* by itself serve to fix any particular conceptual map as the uniquely appropriate map for representing the reality that gets accessed by that modality. There are always different ways of conceiving things, and not just the so-called 'superempirical' things far beyond the reach of our native sense organs. There are also different ways of conceiving the sensorily accessible things to which our sense organs are presumptively responding on a second-by-second basis. Accordingly, if it is underdetermination that poses a problem for Scientific Realism, it is a problem that poses itself right out of the epistemological starting gate. It does not wait until we address the matter of how the sum total of some ontologically 'privileged' set of 'perceptual' judgments might logically underdetermine the truth of the 'theoretical' judgments we might presume to make on their tenuous authority. It begins, as the above list illustrates, with the issue of *which background map* should be the map that our sensory states should be trained, perhaps by our surrounding culture, to index in the first place. It begins with the question of how we should conceive even the *perceivable* world, and it continues with subsequent questions about how we might, and potentially should, *reconceive* that world.

This leaves us confronting, still, a situation where our background conceptual maps are robustly underdetermined by the raw statistics of our preconceptual sensory transductions. But this underdetermination is not *logical*; it is *causal*. Neither the would-be thing determined (the map) nor the would-be determiner (the states of our sensory organs) has a literal truth-value, and neither one bears any logical relations to the other. If there is a problem here, it is that the objective statistics of our raw sensory inputs, all by themselves, do not causally guarantee that any *particular* domain-feature map comes to be the map that gets habitually indexed by our sensory inputs. *This* is the form that underdetermination takes within the cognitive framework on display.

Very well then; what further developmental factors, when conjoined with the raw statistics of our sensory inputs, *do* suffice to causally determine the eventual character of our conceptual maps? Most obviously, there is the profoundly important mechanism by which our myriad synapses are progressively modified in response to the statistical profile of our sensory inputs. This mechanism is apparently the Hebbian learning mechanism discussed earlier.

As we have seen, this mechanism tends to produce very similar conceptual maps across diverse individuals, so long as they are subject to sensory inputs with highly similar statistical profiles. To be sure, the resulting maps are unlikely to be perfectly identical, for a number of reasons. First, in biological brains, any population of neurons is only 'sparsely' connected to the neurons in the next population in the processing hierarchy, and these connections tend to be randomly distributed. This can and does produce differences, across individuals, in the acquired profiles of 'preferred stimuli' that constitute the individual axes of their conceptual maps. The global configuration of 'prototype points' within each individual's map will likely be similar across distinct individuals even so, but since those configurations are sustained by a different family of activation-space *axes*, slight metrical differences across the maps of distinct people are only to be expected. The Hebbian-trained face-recognition network of fig. 3.8. will provide an example of such behavior. Train ten networks of that same general configuration, but with randomly different sparse connections, on the same training set of input photos, and you will get ten roughly conformal but metrically idiosyncratic maps of facial space.

Second, although we humans all confront the same empirical world, no two people have exactly the same histories of sensory stimulation, and so once again we can expect similar, but not identical, maps to emerge from the learning process. Most important of all, cultural differences in the environments people inhabit, and cultural pressures from other people, to pay attention to this, that, or the other aspects of that environment, can lead to interesting and occasionally substantial differences in the maps deployed at different stages of human history. The contrast between the ancient Greek's conception of temperature and the modern Fahrenheit conception is a good example. The latter map unites both hot and cold onto a single, united scale, and it provides a metric to quantitatively compare different temperatures by mapping them all onto the natural numbers. The contrast between the Fahrenheit and Celsius conceptions provides a second example, although this contrast is rather smaller than the first, being limited to the 'grain' of the imposed scale and the location of its two major signposts. The Kelvin scale, however, introduces a significant conceptual novelty by imposing an absolute lower limit on the range of possible temperatures: 0 kelvins or absolute zero. And the Boltzmann and velocity scales mark truly major conceptual reconstruals of what we are all perceiving, namely, the *kinetic energy of molecules* and the *velocity of molecules*, respectively.

These latter two cases illustrate the kind of conceptual epiphanies that constitute what we have been calling Second-Level Learning, namely, the redeployment of some background conceptual map, already at work in some domain or other, into a new and unexpected domain of phenomena. Once discovered, the spread of that discovered redeployment, through the population, will of course be a function of the social environment and the social pressures discussed above. But its original redeployment typically resides in a single individual (in Herr Fahrenheit, for example, or more dramatically, in Copernicus or Newton), the person whose recurrent modulation of his or her ongoing interpretive activities leads him or her to stumble onto that novel deployment of some prior conceptual resources. Most importantly, such singular events are flatly unpredictable, being the expression of the occasionally turbulent transitions, from one stable regime to another, of a highly nonlinear dynamical system: the brain.

We thus have a third major factor in the determination of which conceptual map ends up serving as the default or spontaneous framework for interpreting the inputs from any given sensory modality. Unlike the first two factors, however, this one lacks the fairly strong tendency we noted above toward conceptual conformity across individuals. For one thing, people differ widely in their basic capacity for creative imagination. For another, which particular analogies happen to occur to any given individual will be very much a matter of their idiosyncratic cognitive history and their current intellectual concerns. For a third, which, of the many analogies that fleetingly occur to one, happen to *stick*, as repeatedly plausible and reliably useful analogies, will vary as a function of idiosyncratic cognitive and emotional needs. And for a fourth, which idiosyncratic culture one grows up in, and what stage of its conceptual history one is born into, will have a major causal influence on which conceptual resources become one's default repertoire. In all, the process of Second-Level Learning would seem to display a strong and unpredictable tendency toward cognitive *diversity*, not conformity.

And so human history records. To be sure, social pressures tend to enforce a shared conceptual repertoire within an established culture. But across distinct cultures, and across historical times, we may observe the most diverse and occasionally absurd theoretical conceptions of human nature, of human development, of human illnesses, of the heavens above, of the Earth and the oceans below, of the causes and significance of storms, lightning, earthquakes, and other natural disasters, of the nature of fire, of the basis of morality, of the origins and destiny of the human race, and so forth. A recurring theme, worldwide, is the reading of sundry *purposes* or *agency* into many of the phenomena just listed, a reflection of the overwhelmingly central role that Folk Psychology plays in the conceptual and practical life of any normal human. We naturally try to redeploy, for explanatorily puzzling phenomena, the conceptual maps with which we are most familiar and most comfortable, and our dearly beloved Folk Psychology clearly tops that list. It is no surprise, then, that human history is replete with 'official stories' concerning the aims and commands of sundry gods, the character and purposes of various spirits, and the 'social or political structure' of the cosmos as a whole.

But even within this strong tendency to anthropomorphize everything in the universe, an unpredictable variety of wildly different stories is what history reveals and cultural analysis confirms. From ancient animisms, through diverse polytheisms, to culturally idiosyncratic monotheisms, we find the human imagination run amok in its attempt to find a stable understanding of the cosmos and our own place within it. Consider the high-dimensional doctrinal space that includes, in one quiet corner, the ontologically modest Chinese reverence for the 'spirits of their ancestors' and for the worthy personal characters those ancestors displayed. That same space also contains, in a less restrained and more colorful corner, the ontologically profligate European devotion to an entire celestial 'Kingdom.' (Notice the redeployment here of a now-ancient *political* prototype.) That Kingdom is populated by an Almighty God/King, an only begotten Son, a furtive rebel Satan with a cohort of evil allies, a large coterie of winged Angels, a gallery of Saints, a teeming multitude of saved human Souls playing harps, a lesser number of hopeful individuals yearning to escape the anteroom of Purgatory, and goodness knows how many sorry souls who will burn in agony, for an Eternity, in that Kingdom's Hellacious basement. (To those of us looking in from the outside, from the perspective of modernity, it is an enduring mystery how the clerics of this religion can continue to tell this story with a straight face.)

Scattered throughout that doctrinal space are thousands of other no less fabulous stories, embraced with equal confidence by other cultures, but shaped by little or nothing in the way of empirical evidence and responsible comparative evaluation. Indeed, the historical tendency has been not a convergence, by distinct cultures, toward a common interpretation of the reality that embeds us all. Rather, the trend has been toward repeated doctrinal fissions, with the divided cultures subsequently going their separate and often hostile ways. Christianity was hived from ancient Judaism, and still shares with it the 'Old Testament,' despite a long-standing mutual distrust. Since then, the Roman and Eastern Orthodox churches have split the original Christian community into two solitudes. As if that weren't enough, the Protestant Reformation subsequently split the already-calved Roman domain into two politically and doctrinally distinct cultures. That distinct Protestant culture itself continues to fractionate into ever-smaller and ever- stranger groups, especially in America and Africa. Judaism, too, has been split asunder at least once, as, most famously, has Islam. Factionalization of doctrine, and gradual extinction for some, seems to be the standard pattern, historically. Grand *fusions* of doctrines and cultures are almost nowhere to be found. The process presents itself as a meaningless historical meander, without compass or convergence.

Except, it would seem, in the domain of the natural sciences. Here also, we see an historical plurality of doctrines, but in the last 2,000 years, and especially in the last 500, we see repeated convergences of doctrine and systematic unifications of distinct scientific cultures. The collected institutions of modern science span every continent and every major culture, in contrast to the self-imposed and self-protective isolations of our various religions. And the scientists in any discipline pay keen and constant attention to the theoretical activities and experimental undertakings of their fellow scientists everywhere on the globe, even when, or *especially* when, the 'doctrinal' inclinations of those foreign colleagues differ from their own. Indeed, and once again in curious contrast to the incurious religious clergy, the focused evaluation and ultimate resolution of those differences are what consume most of the professional lives of most research scientists.

To be sure, the outcomes of one and the same experiment can be, and regularly will be, interpreted quite differently from within different theoretical paradigms: there is still no absolutely neutral touchstone of empirical judgment before which all must bow. But distinct interpretive practices can be evaluated for their internal consistency, and then compared to each other for their overall success. Such experimental indexings can also be evaluated for their consistency with distinct experimental indexings made within distinct but partially overlapping interpretive practices, as when one uses distinct measuring instruments and distinct but overlapping conceptual maps to try to 'get at' one and the same phenomenon. That is, one can get a third, a fourth, and a fifth independent voice into what was originally a two-party discussion. Finally, one can compare the contending maps themselves, each as a whole, with the structure of other maps of neighboring, or subordinate, or superordinate domains, with the aim of evaluating, in various dimensions, their comparative conformity or disconformity with those other maps, maps that have already earned some independent claim on our commitment.

By these means, as we have already discussed and as our scientific history illustrates, we can subject our existing conceptual maps to systematic evaluation and, occasionally, to well-motivated modifications. And to judge by the comparative levels of success displayed by successive maps as this evaluative process has been repeatedly applied over historical time, it often leads to major advances in our capacity to anticipate and to control the diverse behaviors of the empirical world. That is, it displays *progress*, sometimes highly dramatic progress, at least in the practical domain.

But does this mean that our successive maps of the universe's timeless structure are gradually converging on a single, 'perfect,' and final map thereof? Almost certainly not. That our sundry overlapping maps frequently enjoy adjustments that bring them into increasing conformity with *one another* (even as their predictive accuracy continues to increase) need not mean that there is some *Ur* map toward which all are tending. The point can be illustrated with various examples from our own scientific history. The Ptolemaic tradition in astronomy pursued a succession of 'planetary-motion' models (those 'planets' included, recall, the Sun and Moon). Each of these models had an updated estimate of the Earth-centered 'orbital' radius of the relevant 'planetary' body, the radius of its rather smaller epicycle, and the placement of such things as the 'equant' points (the point relative to which the objective *angular* displacement of any particular planet was a constant over time).

These successive versions of the general Ptolemaic framework were both progressively more successful in their predictions of the planetary positions as seen from Earth, and successively closer to each other in the proprietary clutch of orbital parameters that each postulated. Ptolemaicists could justly hope that they were closing in on the truth, at least during the early centuries of their quest.

But the long run proved to be different. No one model worked perfectly, and subsequent adjustments designed to improve one dimension of performance also led to serious performance *decreases* in some other dimension (in predicting planetary brightnesses, for example). In the end, the whole thing blew up in their faces, with Copernicus finally replacing the Earth with the Sun as the true focus of all of the planetary motions. The Sun thus ceased to be a planet entirely, and the Earth was demoted (or rather, elevated) to being just one among the many true planets, spinning and revolving around the heavens, like the others, but in its own idiosyncratic ways. Despite the very real convergences displayed in the early stages of that research program, a perfect Ptolemaic *Ur* map of the planetary motions proved not to exist. Accordingly, the successive Ptolemaic models were not converging toward it, their mutual convergence, for a time, notwithstanding.

But the Ptolemaic adventure is a very modest example of the larger point being made here, an example chosen for its brevity and simplicity. A much more dramatic example of a 'mis-aimed' convergence is the history of Classical Mechanics, and the Grand Synthesis of classical physics in which it played such a central role. The reader will recall the story that begins with Galileo's realization (from his inclined-plane experiments) that a moving body genuinely free of all forces will maintain its original velocity indefinitely. Descartes corrected Galileo's well-motivated but false additional assumption that, in the large scale, the path of such inertial motions would be *circular* (Galileo was hoping to account for Copernicus's *planetary* motions as inertial), and he (Descartes) gave us an appropriately *rectilinear* view of unforced motion, a view subsequently embodied in what is now called Newton's First Law. In addition, Newton's own Second Law, relating acceleration, force, and inertial mass, gave us a firm predictive and explanatory grip on *non*uniform velocities and *non*rectilinear trajectories, as are displayed by arrows, cannonballs, and all terrestrial projectiles.

While this explicitly *dynamical* theory was slowly taking form, the bare *kinematics* of the various planetary motions was proceeding quite independently, moving from Copernicus's new system of Sun-centered circles to Kepler's even newer system of ellipses, with the Sun always at one of the planetary ellipse's two foci. Kepler also noticed that each planet sped up as it approached its orbital perigee, and then slowed down again as it approached its apogee, always carving out equal orbital areas in equal times. As well, the orbital *periods* of the planets, located progressively farther from the Sun, increased rigidly with their mean orbital radii, according to the formula $P \propto \sqrt{R^3}$. These three kinematical claims, once again, constitute Kepler's three laws of planetary motion.

Both the dynamical and the kinematical developments count as internally convergent research programs, but those convergences were still feeble compared to the astonishing mutual conformity that subsequently emerged when Newton's Law of Universal Gravitation was added to the overall mix. As we noted earlier in our discussion of intertheoretic reductions, if the presumably super-massive Sun exerted a centripetal *force*, falling off as $1/R^2$, on each and every planet, then it would continuously *deflect* each planet's natural (rectilinear) inertial motion into an . . . ellipse! With the Sun at one focus! And each planet would speed up and slow down so as to carve out equal areas in equal times! And the successive planetary periods would increase as $\sqrt{R^3}$!

Famously, the matured *dynamical* map—originally crafted for terrestrial motions—portrayed the same planetary behaviors as did the matured *kinematical* map, originally crafted for extraterrestrial motions. That is, the two maps gave mutually conformal descriptions in the domain where they overlapped. The mutual conformity was not (quite) *perfect*. The force of earthly gravity on terrestrial projectiles had to be reconceived as *increasing* (slightly) as the projectile approached the ground. And the focus of any planetary ellipse had to be located, not exactly at the Sun's center, but at the *joint center of mass* of the Sun and the planet at issue, a point only marginally different from the Sun's actual center. Here again we have a case where the close, but imperfect, conformity of two maps justly motivates some adjustments to the subordinate map. From the dynamical point of view, one could easily see *where* the dynamically innocent kinematical map was just a tiny bit too simple, and *why* the (very slight) error had been missed by Galileo and Kepler, respectively.

Further exploration of the possible conformities here brought similar successes. The motion of the Moon around the Earth, and that of the Jovian satellites around Jupiter, showed the same pattern. When applied to those cases, the dynamical story agreed nicely with the observed kinematical story. Slight failures of conformity, such as the faintly anomalous behavior of Uranus, motivated a modest modification to the dynamical story (or rather, to the 'initial conditions' assumed in order to deploy it): namely, the postulation of an unknown planet *beyond* Uranus, but exerting a feeble tug on it. A few calculations suggested roughly where it ought to be, and thus directed, telescopic observers discovered Neptune's tiny blue planetary disk the first night they went looking for it.

But there was still more to come, and from a very different domain: the behavior of *gases*, as we noted earlier. If one assumed, for the sake of argument, that any confined gas is made up of a swarm of ballistic particles (submicroscopic analogues of cannonballs)—particles subject, once more, to Newton's laws—then one could deduce the character of their collective behavior. In particular, one could deduce how the pressure they would exert on the walls of their container would *vary* as a function of how fast they were moving, how massive they were, and how many of them there were inside the container. Deploying those laws (here Newton's Third Law finally looms large) plus a few lines of algebraic manipulation quickly yields the conclusion that the pressure-on-the-walls P would be equal to n (the number of particles) times k (Boltzmann's constant) times $mv^2/2$ (their average kinetic energy), all divided by V (the volume of the container). That is, P must equal

$$n \times k \times (mv^2/2) \ / \ V.$$

But this portrayal of the behavior of a confined swarm of ballistic molecules conforms beautifully to the antecedently known Classical Gas Law:

$$P = \mu \times R \times T \ / \ V$$

where μ is the *amount* of gas (although expressed in gram-molecular-weights this time, instead of in the number of molecules n); R is a constant of proportionality (just like k, above); and T (which corresponds to $mv^2/2$ in the dynamical equation) is the gas's *temperature*. Might temperature simply *be* mean molecular kinetic energy? That would explain our striking conformity here, and a great deal else, besides. Once again, Newton's dynamical map proved to be conformal with an independent map of a novel domain, to the considerable vindication of both. To be sure, and as in the astronomical case, there were some minor details to be cleaned up—initially, both the old and new theories were accurate only for so-called 'ideal gases'—but as before, this mop-up work displayed yet further, and still more fine-grained, convergences.

Unifications on such a grand scale—from the submicroscopic to the terrestrial to the astronomical level—do indeed present themselves as cases of theoretical convergence. And so they are. And they are indeed wonderful. But are they convergences toward a single, unique, and ultimate truth? Alas, once again the answer would seem to be "no." Or, at least, "not necessarily." The fact is, the Newtonian dynamical story, including the law of universal gravitation, turned out to be systematically and ostentatiously *false*. To be sure, all of us who learned the two theories that displaced it—Einstein's Special and General Theories of Relativity—were also taught that, even from this new perspective, Newton's laws remained (approximately) true in the limiting case of phenomena that displayed velocities that were very low relative to the velocity of light, or involved only modest gravitational fields. And we were taught, quite correctly, that this (very limited) mutual conformity was a major point in favor of Einstein's new conceptions. But these lessons, while salutary, tend to hide from our attention four strikingly negative points, which are no less deserving of emphasis, especially in the present context.

First, if we consider the full *range* of possible dynamical phenomena—those displaying relative velocities from zero all the way to those near the velocity of light—the predictive performance of the Newtonian framework becomes exponentially pathetic. In this vast space of possibilities, only the thin plane of near-zero velocities displays phenomena as depicted by Newton. Everywhere else they are decidedly non-Newtonian.

Second, even within this thin plane, the reality as represented by Einstein is discontinuously different from that depicted by Newton. While the numerical *values* of the 'lengths,' 'times,' and 'energies' may differ negligibly, from theory to theory, near that limit, the objective properties themselves are very differently *conceived* within each theory. For Newton, those features are *intrinsic properties* of things, and the predicates that express them are one-place predicates. For Einstein, those features are all *relations* between their possessors and some inertial reference frame, and the predicates that express them are all *two*-place predicates. Their respective conceptions of objective reality are so different here, even in the case of low velocities, that the difference shows up in the very syntactic structure of the predicates deployed.

Third, in every case where Newton saw a body moving on a nonrectilinear path owing to the gravitational force exerted by some mass, Einstein bids us see a body, *entirely free of forces*, moving of its own inertia on a geodesic path within a non-Euclidean four-dimensional spacetime. That is, the whole point of Einstein's General Theory is that there is *no such thing* as gravitational *force*. What large masses do is not exert 'forces' on things. Rather, large masses deform the (otherwise Euclidean) geometry of spacetime so that the natural, unforced, inertial paths of local objects within that deformed spacetime all spiral around and toward the 4-D 'world-line' of whatever central mass is responsible for that local geometrical deformation.

Fourth and finally, for Newton, space and time were entirely distinct dimensions of reality. For Einstein, by contrast, the distinction becomes merely context-relative. For him, spacetime itself is a unified 4-D continuum, with no intrinsic division of the kind Newton supposed. Rather, what counts as a timelike dimension and what counts as a spacelike dimension within that continuum will *vary*, for different observers, as a strict function of their relative velocity.

All told then, the large and multidimensional *differences* between the two stories absolutely dwarf the quite real but comparatively minor and perspectivally pinched respects in which they are mutually conformal. That is, the development and articulation of Classical Mechanics, from the sixteenth century through the late nineteenth, is not plausibly viewed as a process aimed at or convergent upon Einsteinian Mechanics. Indeed, the latter exploded almost all of the constituting assumptions of the former, much as Copernicus' picture had previously exploded Ptolemy's.

A slightly later and entirely distinct explosion of a presumptive convergence was displayed in the unfortunate fate of the Grand Trio of classical force laws, the laws for gravitational force, electrical force, and magnetic force, respectively. These all had exactly the same form: an inverse square law where the attractive force was proportional to the product of the two masses, or the two charges, or the two pole strengths, of the objects involved. Algebraically,

$F_G = g \times M_1 \times M_2 / R^2$ (Newton)

$F_E = \varepsilon \times Q_1 \times Q_2 / R^2$ (Coulomb)

$F_M = \mu \times P_1 \times P_2 / R^2$ (Coulomb)

Such an arresting convergence of form in the laws for *all three* of the known forms of action-at-a-distance suggested to everyone that, somehow, the same *sort* of thing must be going on in all three cases (presumably, some sort of dissipation of each force as it was spread ever-more thinly over a Euclidean sphere whose area increased as the square of the distance from the originating mass/charge/pole involved). But Einstein's new account of gravity rudely scotched that pregnant assumption as well. For on his account, gravitational phenomena were uniquely a matter of geometrical deformations in spacetime and involved no forces at all. If that were true, then something else must be going on in the case of electrical and magnetic forces—something else entirely. After all, they can't involve deformations in spacetime as well: that unique privilege has already been assigned to mass.

This three-way convergence was further and even more deeply upset by subsequent developments in quantum field theory, wherein the dynamical interactions between charged particles came to be explained in terms of the *exchange of discrete photons*, rather than in terms of continuous fields of any kind. This conceptual parting of the ways between our theory of gravity and our theory of electromagnetism (and ultimately our theories of all of the distinct 'forces' presumed to be active at the subatomic level) has matured into a famous contemporary conundrum about how to unify, or even to render compatible, the two great theoretical edifices of our time: General Relativity for phenomena at the large scale, and Quantum Theory's 'Standard Model' for particles and forces at the subatomic level.

Once again, we seem to be looking not at a case of convergence toward a single, unified theory. Rather, we are looking at what is currently a Train Wreck, one made all the more fascinating by the striking empirical success and the accumulated epistemological momentum of the two very different trains involved in the collision.

This is not, I hasten to add, to deny that someday we will construct a theory that *does* provide a unified account of these two domains. I rather expect that we will. But these two vitally important and highly successful research programs do not, or do not obviously, constitute a mutually convergent pair. On the contrary, they constitute a mutually *divergent* pair—divergent in the domains they address, in the mathematical resources they deploy, in the portraits they present, and in the prototypes for further developments that they provide. This makes the current situation in physics highly interesting, even electrifying, because something, it would seem, has to give.

Two lessons emerge from this brief historical summary of some of humanity's scientific adventures in Second-Level Learning. The first lesson, noted earlier, is that the process displays a nontrivial and recurrent tendency toward significant *divergences* in the conceptual resources that people and cultures and scientific subcultures happen to deploy. This contrasts, as we also saw, with the natural and robust tendency toward conceptual *convergence* displayed in the distinct, more basic, and less volatile process of First-Level Learning.

However, there is also a second lesson displayed in these examples. Specifically, the process of Second-Level Learning, as practiced by the worldwide *scientific community*, deliberately and somewhat unnaturally attempts to *impose* a requirement of convergence upon our conceptual and doctrinal developments. This shows itself immediately and perhaps most obviously in the practice of what Kuhn has called "normal science," wherein some discipline or subdiscipline pursues the goal of bringing all of the diverse phenomena in some domain under the interpretive umbrella of a single conceptual paradigm. In this protracted process, the success of that paradigm is judged precisely by its growing ability to achieve a *unitary* suite of predictions and retrodictions (i.e., explanations) across the entire domain at issue. That is, the accumulating predictive and explanatory successes are 'convergent' in that they were achieved by deploying *the same* conceptual resources across *distinct* phenomena within the target domain at issue. The formative periods of Ptolemaic astronomy, Newtonian mechanics, Einsteinian mechanics, and Quantum Field Theory all display this pattern, as do all successful research programs throughout the history of science.

This strong developmental tendency, achieved by the artificial *imposition* of interpretive standards deriving primarily from the enfolding scientific community, leads any research program to continue the articulation, and to try to broaden the application, of any successful conceptual redeployment, in hopes of both deepening and expanding the research community's understanding of the domain at issue. It also leads the relevant discipline to be *skeptical* of any alternative or competing conceptual redeployments that fail to display such fertility. And it further leads the discipline to begin to *withdraw* its commitment to the current front-runner, if that favored conceptual resource begins to show diminishing returns on the discipline's conceptual investment—that is, if it encounters the sorts of chronic 'anomalies' addressed by Kuhn: specifically, phenomena within the target domain that steadfastly resist successful assimilation within the currently guiding paradigm.

Lakatos, as we noted earlier, is another contemporary figure who has emphasized the importance of evaluating research *programs* for their comparative performances over extended periods of *time*, rather than evaluating snapshot slices of them (that is, 'theories') against some concurrent base of 'empirical data.' Both thinkers emphasize the dynamical dimension of the scientific process, as opposed to static 'confirmation' or 'refutation' relations between static theories and static data. And both thinkers emphasize the *collective* or *social* nature of the evaluative activities that endlessly shape that dynamical process.

Once again, we are contemplating a picture of the scientific process—we are calling it "Second-Level Learning"—that contrasts sharply with the linguaformal portrait held out to us by the logical empiricist tradition in the philosophy of science. The process of repeatedly fine-tuning and articulating the conceptual resources originally redeployed, so as to repeatedly extend their successful application to new areas of the target domain, can lead us down unpredictable and surprising intellectual paths. Who would have thought, for example, that the natural trajectory of any isolated object orbiting the Sun would be a *flower-like* trajectory (i.e., a continuously *precessing* ellipse, with the Sun at one focus), before Einstein's General Theory predicted such a path? And who would have thought that an oscillating electric charge of sufficient magnitude would generate detectable transoceanic *electromagnetic waves*, before Maxwell's mathematical expression of Faraday's insights predicted the details of their production and propagation? And who would have believed that molecules would chronically interact with one another via *perfectly elastic* collisions (after all, that would violate the classical Second Law of Thermodynamics), before the new Statistical Mechanics used precisely that implausible assumption to help explain such things as the Classical Gas Law and the various instances of unceasing Brownian motion for microscopic particles suspended in a gaseous medium?

These particular articulations, of course, all produced welcome conformities. The known (residual) precession of Mercury's highly elliptical orbit fit Einstein's flower-pattern perfectly. Hertz (across a room) and Marconi (across the North Atlantic) each generated and explicitly detected the transmission of specially coded EM waves. And the assumption of perfect elasticity, for molecular interactions in particular, positively *explained* why the moving molecules of the atmosphere don't gradually fall lifeless to the ground around our ankles, as an atmosphere of swarming ping-pong balls surely would, even in a vacuum.

Other historical articulations of initially successful conceptual redeployments, to be sure, led to prima facie *failures* of conformity. The Newtonian prediction for the deflection of starlight, as it passed by the edge of the Sun during a solar eclipse, was off by a factor of two. Classical EM predictions concerning the behavior of a family of negatively charged electrons orbiting around a positively charged atomic nucleus were absurdly chaotic and unstable, whereas the orbiting electrons themselves displayed a ruthless regimentation, enough to yield the exquisite structure of the Periodic Table of the Elements. And the very successful Caloric Theory of Heat suggested that the measurable heat released by a constant source of friction on a body (such as a drill bit slowly scoring the barrel of a cannon) should eventually start to fade as the finite amount of caloric fluid originally contained in the body inevitably sank toward zero. But the heat flow did not fade over time. Not ever. It was as if the body contained an infinite amount of caloric fluid, an impossibility on anyone's reckoning.

Such outcomes, both positive and negative, are standardly seen as confirmations and refutations, respectively. But such oversimple categorizations make sense mostly in distant retrospect, for they deliberately downplay the inevitable ambiguity, the uncertainty, and the fluidity of the epistemological situation on the occasion of both kinds of outcomes. After all, any 'refutation' can be undercut by subsequent empirical or conceptual developments. Any 'confirmation' can turn out to be a reflection of factors that ultimately have nothing to do with the theory 'confirmed.' But at the time and in the heat of battle, one is not and cannot be certain which of these fates awaits our initial interpretation of the incompletely probative data.

The truth seems to be, instead, that we are all of us carried along on a fluid, continuous, and frequently turbulent river of empirical and theoretical developments. None of those developments is individually decisive in steering our trajectory, but some of them tend to pull us into some particular path downstream; some others tend to deflect us away from other possible paths downstream; and still others tend to steer us into a local eddy, perhaps, where we end up circling in confusion (or in smug dogmatism) for some time before finally escaping to resume our downstream adventures.

However, the sustaining medium of this epistemological flow is not a population of moving water molecules, as in the analogy just employed. It is a population of active neurons, each one influencing the behavior of thousands or millions of others, a population that plies its unique path through the all-up activation-space of the brain's 100 billion neurons. Which regimes of cognitive behavior it settles into in the next ten minutes, or ten days, or ten years, will be determined in part by its sensory inputs within the relevant period, to be sure. But those regimes will also be determined by the brain's purely internal dynamical adventures: by the conceptual redeployments it falls into, and by the smoothly various *partial* reinforcements and disappointments that those various redeployments discover.

Once again we must ask: given the same objective environment to help steer it, and enough time to get past the local noise or turbulence, does this Second-Level Learning process tend to *converge*, across distinct individuals and cultures, on a single and shared regime of conceptual resources and deployments? Well, it certainly *could*, given a *culturally imposed* imperative to try to bring increasingly large portions of our empirical experience (this will include the behaviors of artificial measuring devices) under the explanatory, predictive, and manipulative umbrella of a single system of conceptual resources. Indeed, we have seen this general *sort* of convergence emerge repeatedly from the ongoing activities of the scientific community over sufficient periods of time. The examples, some of them briefly discussed above, are occasionally impressive and were justly celebrated at the time.

But equally often, it seems, those internally convergent research programs led us into what eventually turned out to be 'box canyons.' They led us into regimes of interpretation that turned out to be incapable of bringing all of the phenomena within the relevant domain into smooth and successful comprehension. They led us into regimes that were ultimately exploded and discarded when some new, more successful, and quite different regime of interpretation swept into power. Indeed, such internally convergent research programs can often lead us in two entirely different conceptual directions at the same time, as we noted with the program of General Relativity for the domain of the very large, and the program of Quantum Mechanics for the domain of the very small. Here the culturally imposed imperative toward Unity of Resources is and remains richly satisfied *within* each program, but that imperative is clearly frustrated when we contemplate them as a pair.

Perhaps these occasional hang-ups are wholly natural, entirely to be expected, and quite consistent with the long-term prospect of a genuinely global convergence. After all, Rome wasn't built in a day, and we have no idea of the magnitude of the epistemological task we have taken on.

Perhaps we just need to be patient. We may be looking at what are just fleeting hiccups from the long point of view.

On the other hand, we also have no clear idea of the limitations imposed on the human brain by its (generous but) evidently finite neuronal resources. Is a mere 100 billion neurons *equal* to the task of embodying a conceptual map that is homomorphic with the truly basic categories and invariant structures of the universe as a whole? I have no idea, although surely it is possible that, as presently constituted, we are *not* equal to that task. Perhaps that task requires 100 *trillion* neurons, or even more. But this issue should not be about us and the details of our current constitution. In principle, we can always grow more neurons, or manufacture artificial cortical material to serve as plug-in augmentation of our current resources. The real issue is whether a vector-coding, Hebbian-updating, activation-space-sculpting, concept-redeploying, world-representing, paradigm-expanding system broadly like ours can reasonably hope to get an increasingly accurate grip on the nature of objective reality.

Here, it once again seems to these eyes, there are clear grounds for some nontrivial optimism. After all, we *already* have a strikingly effective grip on large portions of the objective world around us, and we can *already* see, in some detail, how we came to have it. Understanding the dual processes of First-Level Learning and Second-Level Learning, and how they gradually yield a product that represents at least some dimensions of the universe's timeless *abstract* structure, is no different, in principle, from understanding how an ordinary *optical* camera processes incoming light, and how it produces a film image that represents at least some aspects of the ephemeral *concrete* scene at which the camera's lens is aimed. We do not contort ourselves in metaphysical agonies over the representational achievements and limitations of this latter process, even though we know perfectly well that ordinary cameras are blind to many dimensions of the physical reality in front of them. (For example, they are sensitive to but a tiny window in the optical spectrum; the wavelengths of light to which they do respond carry only a tiny portion of the overall physical information about the objects that reflected that light; the details of very small, and very distant, objects are beyond their representational reach; a 2-D image leaves unresolved many 3-D ambiguities that only a second camera, differently placed, can resolve by stereopsis; and so forth.) And where such objective limitations become important to us, we simply set about to circumvent them. We develop infrared films and X-ray films to reach beyond the standard wavelength window. We make false-color photos to highlight scene characteristics of interest. We craft microscopes and telescopes. We take stereo pairs of a given scene and view them simultaneously in a stereoscope. And many other things besides.

Neither, I suggest, should we contort ourselves in metaphysical agonies over the representational achievements and limitations of a functioning brain, for these are no less an empirical matter, to be explored and appreciated by scientific research, than are the virtues and limitations of an optical camera. We already have some appreciation of how, within limits, the process of First-Level Learning can sculpt a conceptual framework that maps at least the gross abstract structure of any sensory domain. And we also have some appreciation of how the process of Second-Level Learning can redeploy some of those acquired frameworks so as to represent abstract domains that are not so directly accessible to our native sensory systems. That our brains show *some* representational successes, in both the native-sensory and the supersensory domains, is no mystery at all.

Moreover, we have been steadily acquiring an appreciation of at least some of the epistemological *limitations* typically displayed by our native cognitive mechanisms, and with these brought to light, we have set about to correct them or somehow circumvent them. As with the optical camera, many of those limitations reside in the quite narrow window onto objective reality embodied in each of our native sensory modalities. These limitations have been, and continue to be, dramatically superseded by the many *artificial* instruments of measurement and detection that our growing technologies have provided. We are no longer limited to what we can see, hear, feel, and smell with the small suite of biological measuring instruments provided us by the narrow evolutionary pressures of a life lived on Earth. The sensory armamentarium of modern science has given us 'eyes' to see into countless new domains. Given this expanded sensory *access* to the world's broader categorical and causal structure, we can and have constructed much more penetrating and far-reaching maps of this initially hidden target, given no more than our native biological mechanisms for map-construction and map-evaluation.

But those native map-forming mechanisms have revealed themselves to be flawed and corruptible in a variety of ways as well. As we saw in our analysis of how typical temporal sequences can be learned by Hebbian updating in suitably recurrent neural networks, that biological process, all by itself, is unable to distinguish between genuinely causal processes and mere pseudo-processes. This failing is familiar to us from the human and animal tendency to infer direct causal connections between two factors from their mere constant conjunction or succession in time, whether accidental or owed to some hidden third factor that happens to be the common cause of both. To penetrate these ambiguities, we must variously intervene in the sequence at issue, or somehow contrive to observe natural interventions therein. We can thereby hope to separate the causal wheat from the noncausal chaff.

We also noted, earlier in this chapter, the chronic human tendency to redeploy *inappropriate* conceptual resources in hopes of making some problematic domain intelligible, and to stick with those redeployments even in the teeth of systematic failures in their cognitive performance, if critical evaluations are attempted at all, as quite regularly they are not. As discussed earlier, this sort of defect shows itself in the widespread tendency for ordinary people to see various *purposes* displayed in natural events. But it is also displayed in the tendency for older scientists to persist in their commitments to a long-familiar explanatory paradigm, even as a new and better one may be taking wing. To circumvent this cognitive weakness, we deploy the *methodological* armamentarium of modern science, above and beyond its growing sensory armamentarium. We insist on the repeated testing of our favored theories, throughout the domain at issue. We deploy techniques for the evaluation of the statistical significance of our experimental results, for the minimization or elimination of noise, and for the identification of the principal components of variation within large data sets. We deploy techniques of experimental design and control for avoiding potential confounds. These include techniques for avoiding "cherry picking" and other forms of experimenter bias, techniques such as "double blind" procedures to avoid these potential perversions. We insist on the critical scrutiny of the alleged results by the rest of the professional community, and on the need to replicate the relevant findings in independent laboratories. Further, we evaluate the proposed theory's consistency and explanatory consilience with the already established corpus of (provisional) theoretical wisdom accumulated over centuries of prior scientific activity, similarly collective in nature and similarly technique-driven. Finally, we try to divine potential consiliences among a population of new and potentially revolutionary theories, theories that might displace our current conventional wisdom.

With these assembled techniques, and we will no doubt learn others, the performance of Plato's camera can be significantly enhanced over its modest capabilities in their absence. Over the centuries, we have been slowly learning How to Do good science—that is, we have been acquiring some *normative* wisdom—in addition to learning about the objective world itself. Accordingly, and as claimed five paragraphs ago, there are indeed grounds for optimism concerning the brain's capacity to get an increasingly broad and penetrating grip on the structure of the larger reality of which it is a part. Its internal representational canvas is not a two-dimensional surface, as with a conventional camera: it is a 100-billion-dimensional neuronal-activation space. And its representational subject-matter is not the ephemeral configuration of local objects: it is the timeless landscape of abstract universals that collectively structure the universe. But otherwise, the brain and the optical camera are in the same business. The brain is more ambitious, no doubt. But it also has more resources.

We must still concede, to be sure, a form of 'underdetermination' of the conceptual maps that we humans end up embracing. The sum total of environmentally induced sensory activity, over the entire human race and over as many centuries as you care to consider, is very unlikely to produce a uniquely inevitable set of conceptual resources with which we are fated to comprehend reality. The overall cognitive system—at both the individual and the social levels—that engages the world is too endlessly volatile, too subtly chaotic, to guarantee such a unique and stable outcome.

But this point need not be interpreted as the *skeptical disappointment* that is typically drawn from news of underdetermination, especially in its linguaformal, logical, or set-theoretical guises. On the contrary, there remain several salient points that counsel confidence and optimism. For one thing, even if the evolving path of human science is indeed the path of a genuinely nonlinear dynamical system, the range of possible developmental paths *need* not be wildly divergent and mutually dissimilar. The Solar System, for example, is also a nonlinear dynamical system, but for most of its history its planetary paths have displayed only modest oscillations within a small range of dynamical possibilities. From year to year, the Earth's orbital motion may never carve out the same path twice—there are too many small perturbations, and they are always changing—but even so, our annual loops around the Sun remain closely similar to each other. The range of possible paths for our developing science may be similarly clustered within a fairly narrow corridor of conceptual evolution, one dictated by the peculiar nature of our cognitive resources and the epistemological environment in which it is constrained to operate. Accordingly, if we were somehow to restart the scientific process that began with the early Greeks, it might well follow a developmental path that would lead to a set of conceptual resources very similar to those we use today—not identical, but very similar.

For a second thing, even if our intellectual adventure is fated to display chaotic discontinuities, such that 're-running' our intellectual history several times from the same point would yield substantially *different* conceptual outcomes each time, this does not mean that those outcomes, however diverse, are not penetrating and faithful representations of objective reality even so. Two maps can differ substantially from each other, and yet still be, both of them, highly accurate maps of the same objective reality. For they may be focusing on distinct aspects or orthogonal dimensions of that shared reality. Reality, after all, is spectacularly complex, and it is asking too much of any given map that it capture *all* of reality (see, e.g., Giere 2006).

A third and final reason for optimism is that the entirely real underdetermination of theory by sensory input, as here contemplated in naturalistic terms, does not have the profoundly skeptical consequence that there exists a specifiable subdomain of reality—the so-called 'superempirical domain'—that is utterly and forever beyond our grasp or ken. Both the syntactic view of theories and the semantic view of theories have suggested to some authors that, for any theory about the nature of reality as a whole, there is always an infinity of very different but nonetheless *empirically equivalent* alternative theories. These are sets of theories each of which entails exactly the same family of 'observation sentences,' on the syntactic view of theories; or sets of theories each of which contains exactly the same 'empirical substructure,' on the semantic view of theories. Accordingly, they continue, we can have no hope of ever choosing from among those alternative theories on genuinely *empirical* grounds. Any choice from among those accounts of 'superempirical reality' will have to be based on merely pragmatic, or aesthetic, or other nonfactual grounds.[7] If these 'nonempirical' criteria are eschewed, the best we can responsibly say of any of those alternative theories is that it is 'empirically adequate.'

This skeptical inference is profoundly unwarranted, even if one embraces one or other of these two accounts of what theories are, for there is no principled *distinction* between the 'empirical' domain and the 'superempirical' domain that is adequate to sustain such a fabulous conclusion—not on the syntactic account, nor on the semantic account. This is not too hard to see, even from within those more traditional perspectives. After all, our native sense organs are themselves just more 'instruments' of measurement and detection, instruments whose systematic responses to the environment are every bit as much in need of ampliative conceptual inter-

---

7. The most prominent contemporary proponent of this position is Bas van Fraassen. See his systematic defense of what he calls "Constructive Empiricism" in *The Scientific Image* (van Fraassen 1980). For a variety of critical commentaries on this book, plus van Fraassen's replies, see also Churchland and Hooker 1985.

pretation and calibration as are the responses of a galvanometer, a magne-tometer, a mass spectrometer, or any other instrument with a prototypical causal connection to some otherwise hidden aspect of reality. As we saw earlier in this chapter, concerning thermal phenomena, even our sponta-neous conceptual responses to our peripheral sensory activities are under-determined by those activities themselves, and can take an endless variety of possible forms, depending on the historical, cultural, or instructional environment of the person involved. This means, note well, that allegedly 'superempirical' criteria—such as unity, simplicity, mutual conformity or consilience, fertility, and pragmatic usefulness—are doomed to play a central role in determining even our *observational* ontology and our spon-taneous *perceptual* judgments. There is no escaping our dependence on these 'systematic' aspects of cognitive activity: there would be nothing worthy of the name "cognition" without them—not even perceptual cognition.

From the naturalistic perspective of the present volume, the arbitrari-ness of the antitheoretical skepticism voiced above is even more obvious. For all theories are conceptual maps and all conceptual maps are theories. What is important, for any map to be taken seriously as a representation of reality, is that somehow or other, however indirectly, it is possible to *index* it. (Otherwise it remains a pure speculation.) Exactly *how* we manage to index it is an entirely secondary matter. A map may be systematically indexed by the activity of our native instruments of measurement and detection—our native sense organs. Or it may be systematically indexed by the activity of new and artificial instruments of measurement and detection—the instrumental armamentarium of modern science. There is no essential difference, in the epistemological *warrant* that may accrue to any conceptual framework, depending on whether the instruments deployed are natural or artificial, on whether their origins were evolution-ary or technological. The warrant is the same in either case.

This observation leaves the reality of underdetermination entirely in place. But it removes the temptation to think that there is some specifiable subdomain of reality—the 'superempirical' domain—that is entirely and permanently beyond epistemic access by human beings. So long as every aspect of reality is somehow in causal interaction with the rest of reality, then every aspect of reality is, in principle, at least *accessible* to critical cognitive activity. Nothing guarantees that we will succeed in getting a grip on any given aspect. But nothing precludes it, either.

5   Third-Level Learning: The Regulation and Amplification of First- and Second-Level Learning through a Growing Network of Cultural Institutions

1   The Role of Language in the Business of Human Cognition

The reader will have noticed, in all of the preceding chapters, a firm skepticism concerning the role or significance of linguaformal structures in the business of both learning and deploying a conceptual framework. This skepticism goes back almost four decades, where it finally found voice in the closing pages of my first book (Churchland 1979), in a chapter that called into question the relevance and the compass of what were there described as "sentential epistemologies." In the intervening period, my skepticism on this point has only expanded and deepened, as the developments—positive and negative—in cognitive psychology, classical AI, and the several neurosciences gave both empirical and theoretical substance to those skeptical worries. As I saw these developments, they signaled the need to jettison the traditional epistemological playing field of accepted or rejected sentences, and the dynamics of logical or probabilistic inference that typically went with it. The preceding chapters of the present book outline what I now take to be the most promising replacements for the elements of that traditional picture, and the reader will by now have a detailed appreciation of the several rationales behind them. This vector-coding, matrix-processing, synapse-modifying, map-constructing, prototype-redeploying alternative picture, quite evidently, has little or nothing to do with linguaformal structures and their rule-governed manipulation.

And yet, the human capacity for language plays an undeniably important role in the human epistemic adventure. No other terrestrial creature comes within light-years of our own epistemic achievements, and no other terrestrial creature commands language. A mere coincidence? We don't think so. But how does the latter provide such a dramatic boost to the former?

It does so in many ways, as we shall see, ways that become more readily visible once we have escaped the crippling delusion that cognition is language-like at its core. The irony here is that the avowedly *pre*linguistic/*sub*linguistic account of cognition explored in the preceding chapters allows us to appreciate, perhaps for the first time, the truly transformative event that was the development of human language. Indeed, it may be the single most important development in the evolutionary history of the entire hominin line.

A language allows its possessor to do a great many things, but perhaps foremost among them is the steering, or the guidance, or the modulation of the cognitive activities of one's fellow speakers. With a singular sentence, one can 'artificially' index any one of their antecedent conceptual maps, without their having any of the sensory activities that would usually bring that about. And with a general sentence, one can help them to amplify or update their background maps rather more quickly than would otherwise be possible. Simply giving people a list of general sentences is a very poor way to create in them an effective conceptual map, as anyone in the teaching profession comes quickly to appreciate. But those generalizations can serve to focus the students' attention on certain elements of their experience, thereby to make some of the regularities it displays more salient. This will enhance the vital Hebbian updating that eventually makes their comprehension unconscious and their apprehension automatic.

More importantly, if such mutual steering of one another's cognitive activities becomes frequent and widespread, then the process of cognition becomes *collective*. It then involves a number of distinct brains—at least a handful, and perhaps many millions—engaged in a *common* endeavor. Their consensual understanding of the world's general structure, of its local and present configuration, of its immediate past and expected future, is then shaped, not by the activities of and the sensory inputs to a single brain, but by the activities of and inputs to a large *number* of different brains, similarly but not identically situated. This increases both the quantity and the reliability of the singular information fed into the overall cognitive process. And, quite independently of this enhanced singular information, it increases the stability and the integrity of the constructive cognitive activities that such singular information stimulates in the group at large. Accordingly, the overall quality of this collective cognitive activity will typically be much higher than the quality displayed in any isolated individual—especially if we consider this contrast in cognitive quality over extended periods of *time*, for then the advantages of collective inputs and consensual evaluation will have had a chance to accumulate.

Most important of all, perhaps, is that under the conditions here contemplated, the process of human learning is no longer limited to the temporal extent of a single human lifetime, and no longer limited by the imaginative reach of a single generation. While the collective cognitive process steadily loses some of its participants to old age, and adds fresh participants in the form of newborns, the collective process itself is now effectively immortal. It faces no inevitable termination. It can continue indefinitely to reap the unfolding conceptual benefits that it makes possible, and to incorporate them into an evolving mode of cognition that is automatically passed on to succeeding generations of participants, to be deployed, reevaluated, and updated by them in turn. We are now looking at a cognitive process that knows no essential limitations, or at least none imposed by time. More colorfully, we are looking at a fire that need never go out, one that may burn ever more brightly as the centuries flow by. And this is evidently what has happened to the human species, and to the human species alone, since our development of language.

Figure 5.1 provides a cartoon schematic of the process at issue, at least in its earliest stages. The enveloping cylinder in its upper half represents the newly developed institution of human language. The pencil-shaped items within its embrace are the temporally extended world-lines of individual brains, pointlike at conception, expanding as they approach adulthood, and decidedly finite in length. Prior to the development of language, they are innocent of the kinds of systematic mutual interactions that language subsequently makes possible. After that development, we see the initiation of the collective cognitive process described two paragraphs ago. Brains are now interacting in ways that were impossible prior to a shared language. And not long after that, we see, in addition, the occasional interactions between the semantic structure of the language itself and the various individuals who use it. For as noted, both the vocabulary and the broadly accepted sentences of that language are sure to evolve over time, as the cognitive activity it sustains expands and deepens the linguistic community's grasp of reality.

Evidently, once a functioning language is in place, the taxonomy of categories that it embraces will help to shape the conceptual frameworks constructed by the learning infants who are fortunate enough to be born into such a community. The objective world itself will continue to be the primary author of each child's conceptual development, no doubt. But the objects and features to which their attention is drawn by the conversations and activities of their community will loom larger in importance than they would have in social isolation. The gross categorical structure of the

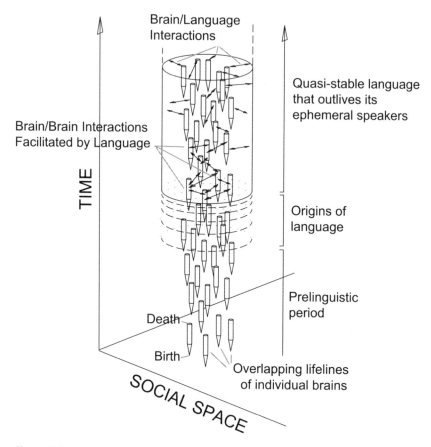

Brain/Language Interactions

Quasi-stable language that outlives its ephemeral speakers

Brain/Brain Interactions Facilitated by Language

TIME

Origins of language

Prelinguistic period

Death

Birth

Overlapping lifelines of individual brains

SOCIAL SPACE

Figure 5.1
The origin of language and (some of) its cognitive consequences.

ambient language thus forms an attractor toward which the development of each child's conceptual framework is drawn.

This gentle assimilation, of the child's developing conceptual resources, to a categorical framework that has already been tested, reconfigured, and found successful by generations of prior cognitive agents, means that each new generation of participants in that activity is the automatic beneficiary of the accumulated successes of that activity. And the inadvertent *victim* of its accumulated *failures*, as well, we must not hesitate to acknowledge. Entrenched nonsense remains ever-possible. But we are here assuming that the benefits of increased informational input and the benefits of consensual criticism and evaluation that characterize the process at every stage will tend, at least on the whole, to produce conceptual maps that are superior to any of those that might have been produced by an isolated individual. Nothing guarantees, as we were forced to acknowledge in the closing pages of the preceding chapter, that the human cognitive adventure will converge on a unique Truth. Dogmatically imposed but misconceived orthodoxies are an ever-present threat to the integrity and long-term fertility of the critical process, as both history and our contemporary culture bear painful witness.

But we also saw that conceptual *progress* remains ever-possible, perhaps even likely, so long as the currently dominant framework is ever-subject to critical evaluation, and so long as alternative frameworks are constantly being invented and explored. Once language is in place, these critical/creative activities can long outlive the ephemeral individuals who sequentially engage in them. Indeed, the forms that those critical and creative activities take are *themselves* subject to augmentation and development, as we are about to see.

2   The Emergence and Significance of Regulatory Mechanisms

Consider figure 5.2, which is itself an elaboration of figure 5.1. It attempts to represent the sequential development of a variety of important human institutions or regulatory mechanisms, all of which are dependent on the prior existence of language, and all of which enhance either the quality of the society's accumulated understanding of the world, or the quality of the cognitive procedures responsible for evaluating and changing it, or both.[1]

Note first that the existence of language makes possible the development of long-lasting oral *traditions*, traditions of instructive stories to be told to children around the fireplace, of enduring legends that inspire awe and virtue in adults, and of religions that maintain a more-or-less systematic family of explanatory doctrines and social practices. With the subsequent development of *written* records, whether on stone, clay, parchment, or paper, the sorts of narratives just mentioned can acquire a stability and longevity that may shape the culture for many centuries. Such quasi-permanent representational media also allow for an accumulating *historical* record that is not entirely fanciful, as the several sorts of narratives just cited typically will be, however benign their original motivations.

---

1. As a formative influence on my own thoughts here, I am compelled to cite the intriguing book by C. A. Hooker (1995), *Reason, Regulation, and Realism: Toward a Regulatory Systems Theory of Reason and Evolutionary Epistemology.*

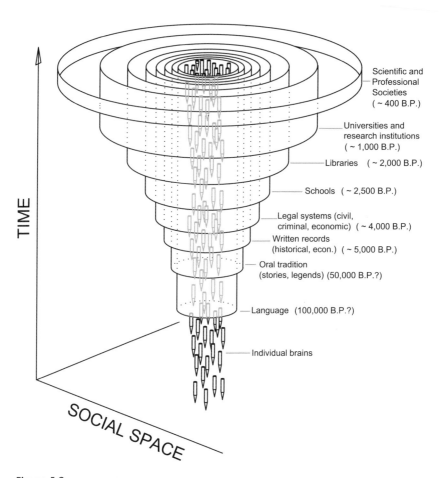

Figure 5.2
The enveloping cornucopia of nested regulatory mechanisms.

Having access to a reliable record of a society's actual past experience, and not just to its current conceptual framework as embodied in its spoken language, gives that society a comparatively stable perspective from which to appreciate the nature of social, political, economic, doctrinal, and conceptual change far more deeply than would be possible with no such record. For we can then evaluate, in retrospect, the actual societal responses that were made to whatever developments confronted it in periods long past, and we can consider possible alternative responses that might have been better made at the time. We then have a chance to learn from our mistakes—not just those mistakes within living memory, but also those permanently inscribed in the growing historical record. We also can come to appreciate the great diversity of *possible* responses to any problem, and the need to explore and examine those diverse possibilities at every stage. As well, a more pedestrian but still transformative function of permanent record-keeping is the *economic* activity that it makes possible. Indeed, most of the surviving clay tablets from the earliest Mesopotamian cultures concern such things as who traded how many bags of grain to whom for how many cattle. Sustaining a stable economic order on any scale would be impossible in the absence of such reliable means of monitoring the exchanges made.

Stable and permanent written records also make possible the institution of systematic civil regulations, criminal law, and economic regulations. In short, they make possible a comprehensive legal system, to shape and govern human activities at a variety of distinct levels. The technology of written records makes it possible not only to monitor our collective activities over time, but to regulate those activities in highly specific and beneficial ways. Explicit written regulations can be posted and enforced much more effectively and uniformly than vague customs and idiosyncratic memories.

Having those regulations written out explicitly also makes it possible to *modify* those regulations in authoritative and highly specific ways, when and if it becomes obvious to everyone that changes need to be made.

Not surprisingly, suitable repositories for the storage, retrieval, and systematic use of these permanent records gradually became common in late antiquity, with the development of sundry libraries and schools devoted to various aspects of our growing sophistication. Institutions such as Pythagoras's mathematical brotherhood in pre-Roman Italy, Plato's Academy and Aristotle's Lyceum in Athens, and the great Library in Alexandria come quickly to mind, and their critical and creative examples were imitated repeatedly throughout the following centuries. By the time we reach the Renaissance in Europe, we see the emergence of long-term institutions of learning and research in England, France, Spain, and Italy, institutions (such as Merton College, Oxford, founded in 1264) that survive to the present day. Collectively, these "universities," as we now call them, constitute an international and multilingual mechanism for the creation, evaluation, and transmission of human knowledge in general, a mechanism that has transformed civilization almost everywhere on the globe.

More recently, the founding of specific scientific and professional *societies*, societies devoted to pursuing research within particular academic disciplines, has created a family of mechanisms for the amplification and regulation of human cognitive activity, mechanisms that are no longer tied to a specific geographical location. The same is true of the proliferation of scientific and professional *journals*. Over time, the administration of both societies and journals can and often does hop from university to university, or it may be permanently distributed across a substantial number of them. The existence of such distributed mechanisms ramps up the temperature of whatever cognitive discipline is their chronic focus, and provides a further means of modulating the cognitive processes of the now-specialized humans who participate in their invaluable activities.

As figure 5.2 suggests, the activities within each layer of this concentric system have an ongoing influence on the activities within all of the other contemporaneous layers, to their mutual and cumulative benefit. And the overall system itself provides an extraordinary environment in which the many individual brains at its core can thrive, cognitively speaking, to a degree that would be impossible in its absence, and at a level that need never stop rising as our collective wisdom accumulates. Overall, it is a perfectly stunning cognitive device, and its ultimate function is precisely the regulation and the amplification of the two basic activities of the biological brains that lie at its core, namely, First-level or *Structural* Learning and Second-level or *Dynamical* Learning. These are both activities that we share with all or most of the other cognitive creatures on the planet, to be sure, but thanks to the unique functional benefits of this enveloping cognitive cornucopia, we have left all of our fellow creatures in the dust. We are not even being chased.

As the diagram portrays, these assembled regulatory mechanisms themselves undergo change over time, and this grand evolutionary process is sufficiently distinct and powerful to merit the appellation of Third-level or *Cultural* Learning. Evidently, we humans have matured dramatically over the last three or four millenniums. We have matured in the basic conceptual frameworks that each new generation is made heir to; in the cutting-edge sciences that we pursue; in the technologies that give us control over so many aspects of the world; in the economic organization that sustains us; in the political organizations that keep our social life stable; and perhaps most important of all, we have matured in the critical methodologies we deploy to *evaluate* proposed changes in or additions to our accumulated regulatory and conceptual capital.

These critical practices show themselves when a journal editor sends a research paper, submitted by some hopeful scientist, out to a handful of mutually independent scholarly referees. Their job is to evaluate the relevance and significance of its alleged results for current scientific activities, the reliability of the experimental techniques used, and the overall quality of the intellectual work done. With these independent experts providing the journal editor with their (possibly diverse) considered judgments, the editor is in a position to make much better judgments about which of the many research papers submitted to him are the most, and the least, deserving of publication and distribution to the journal's many readers. Such judgments are always fallible, of course, but the system just described makes them wiser than they would have been in its absence. Over time, those additional increments of wisdom can make a major difference in the development of the discipline involved.

The same thing happens when scientists submit research papers for potential presentation to a discipline's annual or special-interest conferences, both national and international. The conference organizers, the elected officers of the relevant academic society, use a population of academic referees to advise on the selection of the most worthy submissions, and on the selection of suitably qualified commentators to present informed criticisms of them at the same live session. In the end, the often multitudinous conference attendees (e.g., the recent annual meetings of the Society for Neuroscience in San Diego had some 30,000 registered participants) are treated to critical discussions of some of the best and most recent work in their own field. These discussions can and often do have an impact on those many attendees, who may be inspired or provoked thereby to pursue related researches of their own, either creative or critical. Once again, these organized social practices serve to maximize the distribution of presumptive information, to enhance the quality of its overall evaluation, to expand its applications, and to provoke the construction of novel alternatives by the people affected.

Less obvious, but only slightly less important, is the process by which scholars and scientists apply for research grants to support undertakings that are far too expensive to pursue on their own. The flow of such funds is modest for the humanities, rather larger for the social sciences, and occasionally gargantuan in the medical, biological, and physical sciences, whose research programs often require expensive physical equipment, large numbers of people, and years of concerted effort. As with journal submissions, the applications are evaluated by a suitably expert but presumptively independent group of senior academic referees, people with substantial research experience themselves, people who are the most likely to make wise, if still fallible, selections from among the many proposals before them. The winners of this contest, of course, have only just begun the long process of professional evaluation. For the eventual results of the research thus supported must then be submitted to the journals and to the conference organizers discussed earlier, for evaluation by their referees, and for possible dissemination to the scientific public at large, for critical examination by them in turn, and for the possible modification of their own research activities. That larger public continues to determine the curriculum of our educational institutions, to write the textbooks they use, and to teach their contents to the next generation.

The actors in this worldwide cognitive undertaking are drawn almost exclusively from the global population of university students. These actors are chosen by an extended series of evaluatory mechanisms that determine who gets into graduate school, who is awarded a Ph.D., who gets a job at which research institution, who gets tenure on the strength of past professional performance, who gets promoted to senior positions, and who gets elected to the offices that steer the professional societies, the journals, and the granting agencies discussed earlier. The many and ideologically diverse evaluators, at every stage of this process, are people who have themselves passed through it, and who are continually subject to evaluation themselves, if only by the more elevated stages of that process.

The result is a self-sustaining and self-modifying cognitive organism, whose ultimate function, as we saw, is the regulation, and the amplification, of the cognitive activities at the first two levels of learning, levels located within the biological brains of individual humans. This description may make the overall process seem like a purely intellectual undertaking, but as we saw in the opening pages of this book, cognition is as robustly *practical* as it is theoretical, and the process just described reaches out beyond our universities and research institutions to include the various activities of and professional associations for doctors, lawyers, engineers of all varieties, airline pilots, building and electrical contractors, schoolteachers, and various other information-intensive but primarily practical professions. Over the preceding centuries, the nested mechanisms of figure 5.2 have transformed our practical lives at least as dramatically as they have transformed our theoretical grasp of objective reality. The many practical technologies that sustain the modern world are no less products of the cognitive cornucopia there portrayed than are the theories we embrace and the conceptual frameworks we deploy. Indeed, those nested regulatory mechanisms are ultimately responsible for almost every aspect of modern civilization.

And none of those mechanisms would exist but for language, that original and most basic of all social mechanisms for regulating human cognition. Nevertheless, while our cognition is systematically *regulated* by language—by its contemporaneous internal structure and by the many social mechanisms that it has made possible—cognition *itself* is not linguaformal at all, at least in its primary forms, despite its traditional portrayal as an inference-drawing dance of sundry propositional attitudes. Indeed, language *itself* would not exist but for its acquisition by and its continuing administration by the underlying forms of prelinguistic cognitive activity that we have explored at length in the preceding chapters. Those underlying activities must be addressed, and their basic forms must be understood in detail, if we are ever to understand the ways in which they are so effectively regulated, and so wonderfully amplified, by the hierarchy of social mechanisms that now rest upon them.

3 Some Prior Takes on This Epicerebral Process

As remarked in the opening chapter, the existence of this supercranial process has been well noted and much discussed by a number of earlier authors. There has been little unity, however, in their several portrayals of its essential nature. In an early attempt, Hegel characterized our assembled social mechanisms as the 'World Spirit,' a genuinely engaging metaphor we must surely allow, and he saw its unfolding activities as incremental steps in the slow progress of this Grand Individual toward complete 'Self Consciousness.' Given the plainly cognitive nature of so many of the supercranial activities at issue, and given their increasingly penetrating portrayal of the physical and social universe that sustains them, it is hard not to resonate with Hegel's choice of words here, and hard to resist the grandeur of his vision.

But in the end, the metaphor of a *person* or *mind*, as deployed by Hegel, has little explanatory or predictive substance to it. It does not open any doors for the novel and systematic application of an old concept in a new domain, as the metaphor of a 'flung stone' had done for Newton's Moon and as the metaphor of 'random variation and selective reproduction' had done for Darwin's finches. Indeed, minds are part of our *problem* here, being so mysterious in their own right, rather than a transparent source of illumination for other puzzling phenomena. No significant or progressive research program was ever kindled by Hegel's metaphor, however engaging it may have been, initially. It slipped quietly into explanatory retirement.

A quite different and rather better account of our collective cognitive endeavor was supplied by the Logical Empiricists of the early twentieth century, who portrayed it as a rule-governed process of selecting families of sentences according to their comparative success at predicting and explaining the behavior of the observable world. Interpreting theories or conceptual frameworks as sets or webs of interrelated *sentences* allowed these philosophers to deploy the considerable and highly systematic resources of modern logical theory (which sprang into existence in the late nineteenth and early twentieth centuries) in aid of explicating such notions as *explanation, prediction, refutation, confirmation, intertheoretic reduction,* and *empirical content.* These, in turn, encouraged various accounts of what *scientific rationality* and *intellectual progress* might ultimately consist in. The Logical Empiricists all supposed, to be sure, that the primary locus of this linguaformal activity was inside the heads of the individual scientists involved in our collective scientific enterprise. Nevertheless, they agreed that, through the medium of language, the essential elements of those activities could all be displayed in a *public* cognitive marketplace, there to receive *collective* evaluation by all concerned, according to explicitly stated *rules* of evaluation, rules deriving ultimately from modern *logical* theory. With its elements thus awash in cleansing sunlight, the activities of science will then proceed without the crippling idiosyncrasies and evaluative failings that so often attend the cognitive activities of isolated and unregulated individuals.

Heady stuff. And closer to the mark than Hegel, no doubt. This is the broad philosophical perspective that captured the imaginations of my own generation in the early 1960s, and it steered our research for decades thereafter. Only a few of us have ever escaped it, and many have not yet begun even to doubt it. But doubt it we must. Not because its emphasis on public examination and collective evaluation is misplaced. Not for a moment. Nor because its emphasis on the observational and instrumental interrogation of the empirical world is mistaken. Not in the slightest. We must doubt this perspective, indeed, we must reject it, because theories themselves are not sets of sentences at all. Sentences are just the comparatively low-dimensional public stand-ins that allow us to make rough mutual coordinations of our endlessly idiosyncratic conceptual frameworks or theories, so that we can more effectively apply them and evaluate them.

Nor is the *evaluation* of theories ultimately a matter of determining their diverse logical relations to an imagined set of foundational observation sentences. There is no such set, and the virtue of a theory does not consist in its having any such set of sentences among its specifically logical consequences in any case. Rather, its virtue consists in its being a coherent, stable, and successful vehicle for interpreting and anticipating the perceptual/instrumental experience of the creature who is deploying it. We must remember that the perceptual/instrumental apprehension of empirical reality is *already* a matter that involves the systematic indexing of a *preexisting* conceptual framework or theory, however primitive or comparatively uninstructed. Our conceptualized perceptions—and their ultimate representational adequacy—are an integral part of the *problem* here, not its solution.

To be sure, there is *something* that is conceptually neutral, common to all normal humans, and ultimately crucial to the evaluation of any theories that we may come to entertain or embrace. It is the raw physical activity of our native sense organs. It is the environmentally induced patterns of activity across our epithelial populations of sensory neurons. It is the statistical profile, both spatial and temporal, of the activity at the very first rung of our many-runged neuronal ladders, before that profile receives any processing whatever at the hands of the well-instructed synaptic matrices found at every rung beyond that very first population of sensory transducers.

But we have no public access to those raw sensory activation patterns, as such. Nor do they stand in any logical relations to linguistically formulated theories. For there is nothing remotely linguaformal about them in any case. If you want to be an Empiricist (and this author does), you need a very different story of how sensory experience ultimately steers the conceptual frameworks or theories on whose representational accuracy we depend—especially if you also want to tell a nonfantastical story about empirical cognition within the brains of *animals generally*, because language-like structures seem not to be a feature of the cognition of nonhuman animals at all, let alone to constitute its fundamental framework, as the Logical Empiricists naively supposed.

A better story would examine how, over time and over a variety of cases, the conceptual framework under evaluation performs its job of assimilating diverse empirical examples to the specific categories it contains, of anticipating the further features of such empirical examples as they unfold over time or under broader empirical examination, and of facilitating the creature's control over or manipulation of the empirical examples at issue. Here we return to the metaphor, explored in chapter 3, of navigating a territory by continually indexing whatever presumptive maps one may possess of that territory, and by continually comparing the cognitive performance of distinct maps in those critical areas in which those diverse maps presumptively overlap in some fashion. Cognitive dissonances of various sorts—surprises, disappointments, confusions—can and will motivate modifications in some aspects of the background maps deployed, or the manner in which they get indexed in specific empirical domains. To be sure, in the special case of humans, explicit linguaformal discussion of such epistemic adventures, carried out in public, can facilitate their evaluation dramatically and help to modify our underlying epistemic practices accordingly—but not because those underlying epistemic practices are themselves linguaformal. They aren't. They are vector-transformational. But let us move on.

Some years ago, a radically naturalist portrait of human cognition in its social setting flowed from the pen of Richard Dawkins, sketchily in his early book, *The Selfish Gene* (1976), and in more detail in a later book, *The Extended Phenotype* (1982). This evidently engaging approach is now widely called "memetics," and it draws its explanatory metaphors from Evolutionary Biology and from the behavior of natural replicators such as viruses. It posits social-level replicators called *memes*, as instanced, for example, in a piece of gossip, a bit of news, a catchy pop tune, and most importantly, for our purposes, in sentences (and sets of them) that for some reason tend to command assent in those of us who hear or read them. The idea is that theories in particular, like other sorts of memes, can spread through a population by word of mouth, or print of page, 'infecting,' as it were, increasing numbers of people by lodging themselves—perhaps firmly, perhaps precariously—in the belief systems of the individuals who encounter them. The long-term processes of collective scientific cognition are thus of a piece with evolutionary processes generally, on this view. They display spontaneous variation in, and selective retention and multiplication of, theories of competing types.

And so, perhaps, they do. Whether this metaphor throws significant explanatory light on the process is, however, quite another matter. The dynamical parallels between a virus-type and a theory-type are pretty thin. A token of a given theory-type does not literally replicate itself in the way that a token of a given virus-type certainly does. Nor does it produce millions of identical daughter tokens in the infected individual. Neither do theories typically come into existence in the first place as accidental tiny variations on preexisting theories, as all viruses do. Neither does the metaphor do anything to illuminate the presumptively special features of the complex social dynamic for generating and eventually choosing between competing theories. For all the metaphor tells us about that collective process, scientific theories *might as well be* pieces of gossip, or catchy pop tunes, so far as their special functions and principled evaluations are concerned. None of this is to deny that our collective scientific endeavor is ultimately a natural process. Surely it is. But it is entirely unclear that the metaphor at issue has gotten to its essential nature or to an explanation of its unusually powerful products.

These complaints loom larger when we remind ourselves that, on the view proposed by Dawkins, the basic units of selection here are still supposed to be *sentences*, or *sets* of sentences. The replicating-memes story may well have transcended the stodgy normativity of the Logical Empiricist's oversimplified rule-bound account, but it remains wedded to the dubious idea that *linguaformal* entities are the basic targets of cognitive evaluation and selection, instead of the high-dimensional neuronal *maps* of abstract feature-domains touted throughout this essay. Dawkins' story, though novel and agreeably naturalistic, once again attempts, like so many other accounts before it, to characterize the essential nature of the scientific enterprise without making any reference to the unique kinematics and dynamics of the *brain*.

To reprise an earlier metaphor, that is like trying to characterize the essential nature of *Life* and *Health* without making any reference to the biochemistry of bodily metabolisms, to the molecular genetics of growth and reproduction, and to our immunological mechanisms for repelling viral and bacterial invaders. Society has indeed many highly-developed *social*-level mechanisms—farms, supermarkets, clinics, hospitals—for sustaining life and promoting health, and those mechanisms are precious, to be sure. Clearly, however, no account of life or health that confines itself to those social-level mechanisms alone will ever get to the heart of the matter. For that, the microstructure of the body and the nature of its microactivities are uniquely essential. Similarly, no account of science or rationality that confines itself to social-level mechanisms alone will ever get to the heart of that matter. For that, the microstructure of the brain and the nature of its microactivities are also uniquely essential.

This agreed, the nested regulatory mechanisms of figure 5.2 remain a major part of what has made science possible, and their role in shaping the scientific enterprise is something that commands systematic examination at all levels. In particular, we need an account of their actual structure and of the details of their functioning if we are to understand their effects on the internal cognitive activities of the individual people whose cognition they regulate. We need, that is, to understand the *sociology* of our scientific institutions, and their systematic interactions with the structural (first-level) and dynamical (second-level) brain activities of individual scientists. This recommendation may send a *frisson* of discomfort down some spines, given the uneven quality and openly skeptical character of what has been presented as Sociology of Science in recent decades. But pursue it we must, if only to enhance the *quality* of those regulatory institutions.

Most obviously, the role played by textbooks, teachers, structured curriculums, and functioning laboratory communities in gradually introducing any would-be scientist into the profession at large, and into the skills and ambitions of a specific discipline in particular, must be understood in greater and quite different detail than we currently command. No one currently asks, let alone tries to answer, how these social-level factors gradually shape the neuronally embodied cognitive maps that actually constitute the budding scientist's conceptual framework and acquired practical skills. No less important, the role played by creative imagination, critical exploration, collective evaluation, and consensus formation in the business of map *re*deployment and *second*-level learning must also be understood in greater, and probably quite different, detail than is currently available. One can perhaps forgive a generation of sociologists for being skeptical about the rationales offered by the scientists themselves in putative explanation of the methodological decisions they embrace and the theory choices they make as the scientific enterprise proceeds. For the scientists themselves may indeed be confabulating their explanations within a methodological framework that positively misrepresents the real causal factors and the real dynamics of their own cognitive behaviors.

One can appreciate this possibility immediately by imagining that the *Hegelian* framework had been embraced wholesale by the late-nineteenth-century scientific community, as the authoritative account of the kinematics and dynamics of scientific cognition, so that every scientific development came to be explained in terms of the interactions among a *thesis*, an *antithesis,* and a subsequent *synthesis*, which brought us, collectively, closer to a complete *self-consciousness*. (This possibility may require a fair bit of creative imagination, but try to enjoy its exploration, if only cynically.) As it happens, and very fortunately, the Hegelian framework was *not* embraced by the scientific community as the default framework for constructing specific scientific rationales. That distinction fell to the Logical Empiricist framework, in which all scientific developments were explained in terms of the *logical* and *mathematical* relations holding among various *theoretical hypotheses* and sundry *observation sentences*, a process of evaluation that was thought to bring us closer to an *ultimate truth*.

This was, I am confident, a much happier historical development than the Hegelian scenario just imagined. But you can readily appreciate at least the abstract possibility that *this framework also* misrepresents the real kinematics and dynamics of scientific cognition, especially in light of what we have learned about that process in the preceding chapters of this book. This possibility may seem to justify the principled decision made by many sociological investigators to 'bracket off' the explanatory rationales provided by the scientists themselves (in explanation of their doctrinal choices and epistemic behaviors), and to replace those suspect explanations with explanations that appeal solely to the obvious social-level pressures and forces—such as professional authority, the desire for peer-group approval, career advancement, and continued grant monies.

But it doesn't justify any such thing. Whatever the ultimate status of the broadly Logical Empiricist framework that is admittedly presupposed by the evaluative activities of contemporary scientists, that framework remains a central and causally potent element in the dialectical and decision-making activities of the scientific community. In virtue of being widely and uncritically *accepted* by the scientific community, that framework plays a major *causal role* in the evaluative and decision-making activities of that community. To elide, deny, or ignore those activities, and the forces that shape them, is therefore to forgo a wealth of explanatorily relevant materials and to replace them with the crudest of political caricatures.

A relevant parallel lies in the question of how to explain the voting behaviors of citizens at the voting station, or legislators in the Senate, as they try to pass or reject a substantive legal initiative or piece of legislation. For better or for worse, the *moral* convictions of those agents will play a major role in determining their voting behavior. To be sure, one may be deeply *skeptical* of the moral convictions of the citizens, or the senators, involved. Indeed, one may reject those convictions entirely, on the grounds that they presuppose some irrational religion, for example. But it would be foolish to make a policy of systematically *ignoring* those assembled moral convictions (even if they are dubious), if one wants to understand the voting behavior of the individuals involved.

A second parallel is displayed in the design and construction behavior of early nineteenth-century engineers, as they busied themselves to create, let us suppose, a large railroad locomotive powered by steam. The dominant thermodynamics at the time portrayed heat as a subtle substance called *caloric fluid*, and thanks to the theorizing of the French scientist Sadi Carnot (*Reflections on the Motive Power of Fire*), our understanding of the behavior of steam engines depended profoundly on that (false) theory. Nevertheless, that theory guided the behavior of the engineers cited, and any attempt to understand and explain their working behavior would depend crucially on appreciating the details of that background framework, false though it was.

The activities of scientists are no different. To understand *their* working behavior requires that we understand the rationales they give to one another, even if the background framework in which those rationales are expressed is in some way faulty or superficial, a possibility that is quite real. This will require that our investigations into these regulatory frameworks be conducted by people who are intimately familiar with the particular scientific discipline under study—with at least its recent history, with its conceptual and instrumental resources, with its current problems and explanatory ambitions, and with the virtues and vices of its contending theories, as conceived by the people within that discipline. This asks a great deal of any would-be investigator, to be sure, but to the extent that you are not capable of being a competent and fully embedded *participant* in the activities, arguments, and decisions of the research community under study, you are unlikely to appreciate the substance of what is going on, or grasp the real causal factors that shape how that community's research develops.

The fact is, the assembled social institutions that shape modern science are the principal *reason* for its stunning explanatory, predictive, and technological successes. They guide its intricate activities in much the same ways that our elected legislatures, our civil and economic regulations, and our judiciaries guide the collective activities of a modern society. As regulatory mechanisms, they are no doubt imperfect, but over time, their shortcomings can become evident, and they, too, can be amended. Just as our scientific theories themselves can become increasingly accurate, penetrating, and comprehensive, so too can our scientific *methodology* become wiser and more effective. Figure 5.2 is itself a crude picture of that very process unfolding. And no doubt additional regulatory rings will be added to that picture in the centuries to come. Human science has only just begun.

4   How Social-Level Institutions Steer Second-Level Learning

Putting aside the obvious and massive advantages that the nested institutions of figure 5.2 provide for the process of *first*-level learning, how do any of them help to steer, or to modulate, the more mercurial processes involved in *second*-level learning? That is, how do they help us to recognize the *need to formulate* new hypotheses or new forms of understanding in a given domain? How do they help us to *come up* with them, and how do they help us to *evaluate* such proposals after they have been formulated?

Most obviously and as already noted, a shared language and a shared theoretical vocabulary allows us to *coordinate* our individual conceptual maps of a given domain, both in our background understanding of that domain's abstract structure and in our local sensory and instrumental *indexings* of that map in the business of interrogating the objective world. Being hidden inside the brain, and being at least slightly idiosyncratic to each person, those complex maps need to be mapped onto *each other* by some means if their possessors are to engage in *joint* cognitive activity. A shared theoretical lexicon, and a shared family of accepted sentences expressed in that lexicon, makes such interpersonal map-coordination possible, if only in gross outline. The extraordinary intricacy displayed by individual maps with hundreds of thousands or even millions of distinct and idiosyncratic neuronal dimensions will likely preclude any *perfect* mutual mappings or interpersonal coordinations, as we have already seen. No two people will conceive of the world in *exactly* the same way, even if they learned their theories in the same classroom from the same teacher. But rough mutual mappings will still be entirely possible. And once achieved, they can be profoundly useful, for then the ensuing cognitive activities of the people thus coordinated can serve as a critical check on each other if they happen to diverge, or as a confirmation of each other if they do not diverge.

Interpersonal differences aside, such rough coordination of the conceptual maps of the members of a scientific community, and of their practices of indexing those maps (perceptually and instrumentally), makes it possible for the entire community to recognize when, and roughly where, their attempts to 'navigate' the domain at issue lead to perceptual or instrumental results that are systematically at odds with what their shared map so far leads them to expect in that domain. In other words, it makes it possible for that coordinated community to recognize what Kuhn has called *anomalies* in the behavior of the domain supposedly portrayed by their shared conceptual map. Isolated hiccups and puzzlements are not the primary issue here, although a sympathetic coworker can certainly help one to reintegrate a recalcitrant result into one's settled understanding (as in "Try it again with that electrical connection tightened," or "Retake the picture with the lens-cap *off* this time"). Rather, the primary issue here is the emergence of experimental results that repeatedly defy the expectations of an entire community despite the continued attempts of its members to fine-tune the experimental techniques used, to reassess the background assumptions made in applying them, or to reinterpret the background conceptual map being deployed, so as somehow to restore and maintain a smooth equilibrium between their map-induced singular expectations and their actual perceptual or instrumental results.

Granted, such cases may differ only in degree from the humdrum cases just put aside. But the magnitude, stubbornness, and universality of large-scale anomalies can serve to focus the attention of the entire disciplinary community, and to foster extended efforts to repair the increasingly frustrating cognitive situation. Such efforts can range from the conservative and ameliatory, to the modestly but creatively reconstructive, and to the downright revolutionary, as in those comparatively rare cases when the community's confidence falters, the old conceptual map comes increasingly to be seen as fundamentally inadequate to the domain at issue, and attempts are made to deploy a new and quite different map in its place. As we have seen, Kuhn describes such periods of turmoil as 'crisis science,' and he explains in some illustrative detail how the normal pursuit of scientific inquiry is rather well-designed to produce such crises, sooner or later. I am compelled to agree with his portrayal, for, on the present account, 'normal science,' as discussed at length by Kuhn, just *is* the process of trying to navigate new territory under the guidance of an existing map, however rough, and of trying to articulate its often vague outlines and to fill in its missing details as the exploration proceeds. If that map is indeed a fundamental misrepresentation of the domain at issue, such *collective* and *mutually coordinated* explorations conducted under its presumed authority are precisely what is needed to gradually determine that, and where, the map fails of its purpose.

Significantly, from the perspective of the present book, Kuhn does *not* speak of sentence-like hypotheses being logically contradicted, in the context of sentence-like background assumptions, by sentence-like observation-claims. Instead, he speaks of 'guiding paradigms' and of the frustrations and disappointments occasionally, even regularly, encountered in trying to expand their applications into modestly new territory. Like Lakatos, Kuhn sees 'normal science' as embodying a research *program* that moves erratically forward in a sea of *ever-present* anomalies, both minor and major, anomalies whose significance is rarely clear at their first encounter, and whose resolution often takes time, sometimes a great deal of time. This overall process, moderately well-described by both authors, is of course the site of intricate linguistic activity on the part of the scientists involved. And a good thing, too. For language facilitates alternative articulations of the fuzzy background map, well-focused applications of them to the problematic domain at issue, and consensus evaluations of their empirical success or failure. But the real kinematics underlying this invaluable chatter remains the business of trying to deploy a very high-dimensional and only partially articulated map to an ambiguous and only partially appreciated empirical feature-domain, in hopes of having the *expectations* thereby generated either *rewarded* or *disappointed*.

The outcome of this business, whether successful or unsuccessful, will no doubt be reconstructed by its actors in logico-linguistic terms, as a hard-earned *confirmation* of the map's basic integrity, or as a presumptive *refutation* of either the map or of its peculiar application to the present case. And such reconstructions can be highly useful. But they always exaggerate the degree to which the overall situation is explicitly understood, because they use terms whose meaning is already embedded in a complex, fluid, and not entirely determinate network of background assumptions, which assumptions are often themselves part of what is at stake in the research program at issue. As well, those logico-linguistic reconstructions portray the cognitive situation in terms of *discrete* categories and *binary* (true/false) evaluations: that is, they provide a 'digital' representation of what is a genuinely continuous and profoundly 'analog' process of representation within the brains of the researchers involved. Those procrustean reconstructions are an invaluable part of the cognitive process, and they play a genuine causal role. But we should not presume that they embody the true cognitive kinematics and the real cognitive dynamics embodied in the vector-processing activities of those very same speakers and writers. They don't.

One can appreciate this point from a different perspective—that of our recorded scientific history. When one reads the original works of prominent scientists working prior to the late nineteenth and early twentieth centuries—scientists such Aristotle, Bacon, Galileo, Descartes, Leibniz, Newton, Herschel, Davy, Faraday, and Darwin—one does *not* find them using the kinematical and evaluative vocabulary so common among today's scientific researchers and philosophers of science. One looks in vain for discussions of 'general hypotheses,' of 'initial conditions,' of 'observational predictions' and their 'deduction' from distinct elements of the first two kinds. One finds, instead, talk of 'investigations,' 'inductions,' 'analogies,' 'explorations,' 'phenomena,' and 'consiliencies.' One finds the occasional reference to 'hypotheses' primarily to disdain them, as in Newton's famous "*Hypotheses non fingo*" ("I do not make hypotheses").

The more modern and more familiar modes of scientific self-descriptions do not fully emerge until the late nineteenth century, presumably as a reflection of the increasing *mathematization* of the more successful sciences, in which general laws and formal deductions drawn therefrom played an increasingly prominent role. The philosophical movement subsequently known as Logical Empiricism, accordingly, was simply following a narrative trend already underway in the sciences themselves, independently of the 'linguistic turn' in Philosophy and the maturation of modern Logical Theory at the hands of Frege, Russell, and others. The scientists themselves were simply getting better and better at mutually coordinating their individual conceptual maps, at coordinating their tentative applications to diverse phenomena, and at achieving consensus concerning the success or failure of those focused cognitive forays.

That improved coordination was not uniform across the sciences, however. For the special sciences—such as geology, biology, psychology, and evolutionary theory—the Logical Empiricist framework was a much less welcoming canvas with which to capture their dynamical profile. One still hears it said, often critically, that "there are no genuine laws" in any of these sciences. And the still-iconic Karl Popper retained, late into his life, a systematic suspicion of Darwin's evolutionary theory in particular, based on the presumption that it failed his particular test of being 'empirically falsifiable.' The present author suggests that these particular criticisms of the sciences mentioned are ill placed, and that whatever substance they contain reflects the procrustean nature of the familiar Logical Empiricist framework more than it reflects any intrinsic defects in these sciences themselves. Not all conceptual maps display the spare simplicity of classical mechanics, and not all domains admit so readily to idealization and (over)simplification as do the motions of our planets around the Sun. But those maps all guide our attempts to get a revealing and more detailed grip on the domains they purport to portray. And language makes it possible for those attempts to be both collective and cumulative.

If language, and the social institutions that it supports, make it possible for major cognitive crises to be recognized and isolated, what role does language play in generating the invaluable cognitive epiphanies or conceptual redeployments that so often lead to the revolutionary *resolution* of those crises? Comparatively little, I suggest, beyond (of course) motivating the search for them. Here, even the Logical Empiricist tradition concurs with this negative estimation, at least roughly. That tradition claimed to provide systematic illumination for what was often called "the context of evaluation and justification," but it made no such claims for "the context of *discovery*." Indeed, that domain was typically pushed aside as a matter for *empirical psychology*, as opposed to *normative epistemology*. This (often dismissive) stance toward a naturalistic account of scientific cognition was reasonable enough, perhaps, given that the logical/linguaformal framework then in play had no visible resources for even addressing the matter of cognitive creativity. It still doesn't.

By contrast, the framework of prototype activations in recurrent neural networks provides an immediate characterization of both those cognitive epiphanies themselves, and their subsequent evaluation as potential guiding paradigms for a novel research program in an already existing but troubled explanatory domain. In chapter 4, we explored this process from a neurocomputational perspective, and that same perspective provides a natural explanation for the characteristic features of such 'insightful' episodes: why they have such a short time scale, why they tend to occur in people who possess a large library of comparatively detailed maps ready for potential redeployment in the first place, and why they tend to occur in people who are especially skilled in the recurrent modulation of their perceptual activities and their interpretive habits. (Such people are simply the more creative folks among us, for that is presumably what creativity consists in.) To be sure, nothing guarantees that any one of these mercurial conceptual redeployments will have any objective representational merit. Indeed, most will fade away as brief and unproductive metaphors. But some will stick, subsequently to be spread (via language, etc.) to the rest of the scientific community. If they are judged, therein, to have merit, they can subsequently spark a new program of ongoing research and engage the critical machinery of the entire scientific discipline.

An important source of such revealing 'metaphors' has gone unmentioned to this point, namely, the domain of modern Mathematics. In particular, the entire range of possible *relations* and *functions* provides an infinity of possible models with which one might hope to comprehend the relations that structure some problematic feature-domain. So far in this book, I have spoken of the redeployment of antecedently possessed conceptual maps as if those maps were invariably drawn from our empirical experience, as indeed so many of them were, especially in the earlier stages of our scientific adventure. But there is no need to acquiesce in that restriction, and every reason to step beyond it. It would be miraculous if the narrow *perceptual experience* of we cloistered and blinkered humans should happen to display all of the abstract structures necessary to get a cognitive grip on any and all of the complex feature-domains that constitute the objective physical world. For many vitally important domains are simply beyond the reach of our native sense organs—the subatomic and supergalactic, the high-energy and low-energy, the short-wavelength and long-wavelength, the very fast and very slow. Fortunately, mathematics allows us to *generate* an endless variety of relational and functional models, far beyond those that might happen to be learnably displayed in our narrow empirical environment.

Accordingly, and over the centuries, the various sciences have success-fully deployed such things as the algebra of irrational numbers, the dif-ferential and integral calculus, non-Euclidean geometries, high-dimensional geometries, the algebra of complex numbers, vector/matrix algebras, and nonclassical operators. These map-making resources provide models that are progressively more alien, perhaps, at least from the perspective of the kinds of maps that constitute our humdrum common sense. But by the same token, they provide a much broader range of redeployable resources than we would otherwise command. And this gives us a much greater chance of finding a successful fit with the less accessible but presumably more fundamental dimensions of objective reality. General Relativity and Quantum Mechanics are the most celebrated examples of these nonclassi-cal deployments, but they are not alone.

In fact, the model of cognition on display in this book provides a further example. The framework of high-dimensional neuronal activation-vectors being 'multiplied' by a 'matrix' of synaptic 'coefficients' to yield new activation-vectors across the neuronal population to which those synapses connect is a novel deployment of some well-known *mathematical* resources. But it is hardly a deployment that draws on elements of our familiar experi-ence. Instead, that unfamiliar framework is locked in a struggle with an old framework that *does* draw extensively on some elements from our familiar experience, namely, the framework of Folk Psychology, an entrenched redeployment that models our inner cognition on the model of our external speech. Cognitive inertia can be an enormous barrier to conceptual change, and it most surely has been in the case here at issue. Perhaps the main purpose of this book has been to weaken that inertia and to break down the relevant barriers. For to these eyes, both the time for change and the opportunity to effect it have now arrived. We have started constructing the new framework.

5   Situated Cognition and Cognitive Theory

A further look at figure 5.2 will likely invoke, in many readers, thoughts of so-called 'situated' or 'embedded' cognition, as recently explored by both social scientists and philosophers (see esp. Hutchins 1995; Clark 1998, 2003). The central idea is that many important forms of cognitive activity are not confined to the brain alone, but reach out to include, as an integral part of the cognitive activity involved, various forms of external 'scaffolding,' such as pencil and paper, slide rules, electronic calculators, dividers/compasses, drafting equipment, instruction books, human interlocutors, and external mathematical manipulations of all kinds. A complete story of the nature of cognition, accordingly, must take all such 'extensions' of our native cognitive equipment into proper account.

And so it must. Even the oversimplified cartoon of figure 5.2 portrays graphically how utterly embedded in a system of regulatory and enabling mechanisms are the brains of modern humans. And what we have been calling Third-Level or Cultural learning is precisely the emergence and development of the various forms of cognitive scaffolding, external to the individual brain, that provide the diverse forms and dimensions of regulation involved in the unprecedented cognitive activities of modern humans. The champions of situated cognition are not wildly inflating a minor feature of human cognition, as one might at first have supposed. (I must confess that this was my own initial reaction, years ago.) Language *itself* is the first and perhaps the most transformative of these scaffolds or *regulatory mechanisms*, as I have been referring to them, and many more comparably transformative mechanisms rest on the broad shoulders of this primary institution. These are not minor features of our situation. As well, without the cognitive environment that these nested regulatory mechanisms provide, the occasional flowers produced by *Second-Level* learning might never have bloomed in the first place, or would be doomed to fade away, unevaluated by the appropriate scientific discipline, unrecorded by history, and invisible to the bulk of humanity. The 'situation' in which one's individual cognition is embedded is plainly essential to the many achievements of which we are all so proud.

But what sort of general account would be *adequate* to the diverse dimensions of that embedding 'situation,' and to the remarkable amplification of human cognition that it apparently provides? Is there any remotely analogous phenomenon that might throw some light—any light—on this apparently singular process? I believe there is, and I shall bring this book to a close by briefly exploring one salient possibility, a possibility that is staring us in the face, for, once more, it involves ourselves.

We not only think: we are alive. We not only cognize: we metabolize. That metabolic activity,[2] as modern biochemistry has been teaching us, is extraordinarily complex, although it is basically uniform across all terrestrial animals. It is also confined primarily to the internal milieu of any

2. Defined by *The Oxford Dictionary of Science* (Oxford: Oxford University Press, 1999) as "The sum of the chemical reactions that occur within living organisms."

animal's body. That is where food is digested, temperature is regulated, proteins and other chemicals are produced, invading bacteria and viruses are attacked and destroyed, damage is repaired and growth regulated, reproduction is conceived, canceled, or brought to term, and so forth. All of us are situated in the physical world, of course. But the mechanisms that subserve any animal's metabolism are located primarily inside its skin.

Even so, a few creatures use *external* resources of a modest sort that help to regulate their internal metabolic activities. Think of the squirrel with a safe haven lined with dry leaves and fluffy moss deep inside an old tree trunk. This extra-bodily arrangement certainly helps to regulate the squirrel's bodily temperature every night, and most critically during the cold winter months. As well, her stash of nuts, accumulated during the fall, allows her to maintain her nutritional intake through the barren winter months, that is, to regulate it across the seasonal variations in nut supply. All of this serves to increase her reproductive potential. It also helps her to see her vulnerable offspring through the earliest stages of *their* metabolic adventures. Our squirrel is deploying some 'extra-bodily metabolic scaffolding' to carry some of the complex burden of staying alive and successfully reproducing. Thanks to that scaffolding, her purely internal resources have been importantly enhanced.

One can cite many other examples of artificially 'situated metabolism,' no doubt. But most of them will be similarly modest—until we address the case of humans. Then we confront a major discontinuity between us and all of the other terrestrial creatures, indeed, a broad spectrum of discontinuities. For example, to help regulate our own temperature, humans have been using *fire* for at least 100,000 years, and our hominin predecessors for longer than that. We also make artificial *clothing*—animal skins, at first—to serve the same function. No other creature on the planet does that. In addition, cooking over a fire also serves to make almost any *food* safer, more easily digestible, and more metabolizable, especially meat of any kind, which gives us a major nutritional advantage over the other creatures, none of which commands fire at all, let alone puts it to this important metabolic purpose.

The gap between us and other creatures continues to widen when we consider the various *vaccines* that we now routinely deploy to enhance our immune systems against a variety of familiar but often deadly bacterial or viral invaders. To these we must add the entire spectrum of post-illness *medicines*—antibiotics head the list—designed to attack or disrupt those invaders independently of our own immune systems. External invaders aside, think of the many metabolic failures to which the human body is intrinsically susceptible, such as diabetes, cancer, and a host of auto-immune diseases, all of which are regularly addressed by suitable *drugs* aimed at the relevant dimension of metabolic activity. And note also the extensive institutions necessary to administer these regulatory activities, institutions such as the medical profession, the biochemical research departments in our universities, the Federal Drug Administration to vet the results of that research, the chemical companies to manufacture the relevant substances, and the hospitals and clinics in which they are finally delivered to the troubled metabolisms they will help to reequilibrate. This 'metabolic scaffolding' constitutes a major portion of our national economy.

And let us not forget the squirrel's humble stash of nuts. We humans not only maintain a gigantic industry to accumulate and distribute a stable supply of food to the entire population; we sustain an even larger industry to produce and harvest it in the first place: agriculture. As one flies over the heartland of the United States or Canada and surveys the checkerboard of cultivated fields marching to the horizon in every direction, one begins to gain an appreciation of exactly how 'situated' is the metabolism of every person on the continent. Without that enveloping 'situation,' most of us would starve. And without the houses in which we live and the buildings in which we work, most of us would perish from exposure, at least in the winters—or from disease of some kind, if we managed to dodge these more elementary disasters. All told, the metabolisms of humans are wrapped in the benign embrace of an interlocking system of mechanisms that help to sustain, regulate, and amplify their (one hopes) *healthy* activities, just as the cognitive organs of humans are wrapped in the benign embrace of an interlocking system of mechanisms that help to sustain, regulate, and amplify their (one hopes) *rational* activities. But the systems that minister to our metabolic activities are even more extensive, by at least an order of magnitude, than those that minister to our cognitive activities. Between them, they constitute most of the machinery of modern civilization. And only humans possess them.

At this point, a romantic (or cynical) reader might suppose that I am here trying to resurrect some novel form of Hegelianism—that is, to portray the organized machinery that sustains our metabolisms as some kind of giant *living organism*, and to portray the organized machinery that sustains our cognition as some kind of giant *mind*. But that is emphatically not my purpose, for two reasons. First, I do not believe that our classical, presci-entific conception of a "living animal" will throw any light at all on the nature of those supra-individual metabolism-regulating mechanisms. And second, I do not believe that our classical, folk-psychological conception of a "mind" will throw any light at all on the nature of our supra-individual cognition-regulating mechanisms.

Instead, what we need in the first case, and already possess, is a *new and scientifically informed* conception of the complex metabolic activities that make any animal a living thing. This has allowed us to appreciate in greater detail the regulatory mechanisms that humans have long employed. It has allowed us to appreciate how they do their nourishing and reequilibrating work, and it has allowed us to generate new regulatory practices on a continuing basis. Similarly, what we need in the second case is a *new and scientifically informed* conception of the cognitive activities that make any animal a thinking thing. This is something that Cognitive Neurobiology has recently begun to provide us, as I have tried to sketch in the pages of this book. And this will allow us to appreciate in greater detail the regulatory institutions that humans have long deployed. It will allow us to appreciate how they do their regulatory and evaluative work, and even, perhaps, to generate new mechanisms—think of computer technologies and the Internet. Old myths and folk conceptions are not what we need at this point, and especially not for redeployment as Hegelian metaphors. Where cognitive theory is concerned, what we need is a comprehensive and revealing theory of *brain activity*. Then, and only then, will we be able to understand in detail the very considerable role that our secondary social institutions play in regulating and amplifying it.

Appendix

Kepler's Third Law Deduced from Newtonian Mechanics

F is force. g is the gravitational constant. M is the Sun's mass. m is the planet's mass. R is the planet's orbital radius. a is acceleration. v is the planet's orbital velocity.

(1) $F = \dfrac{gMm}{R^2}$ Newton's Law of Gravitation

(2) $F = ma$ Newton's Second Law of Motion

(3) $a = \dfrac{v^2}{R}$ Centripetal acceleration of a body in circular motion

(4) $P = \dfrac{2\pi R}{v}$ Definition of Planetary period

(5) $\dfrac{gMm}{R^2} = ma$ from (1) and (2)

(6) $\dfrac{gMm}{R^2} = \dfrac{mv^2}{R}$ from (3) and (5)

(7) $v = \dfrac{2\pi R}{P}$ from (4)

(8) $gMm/R^2 = \dfrac{m(2\pi R/P)^2}{R}$ from (6) and (7)

(9) $gMm/R^2 = \dfrac{m(4\pi^2 R^2/P^2)}{R}$ from (8) (Now, multiply both sides of (9) by R^2, and both sides by P^2.)

(10) $gMmP^2 = m(4\pi^2 R^3)$ from (9)

(11) $P^2 \propto R^3$ from (10) (All elements in (10), other than P^2 and R^3 are *constants*, so they can be deleted.)

(12) $P \propto \sqrt{R^3}$ from (11)

References

Akins, K. 2001. More than mere coloring: A dialog between philosophy and neuro-science on the nature of spectral vision. In *Carving Our Destiny*, ed. S. M. Fitzpatrick and J. T. Bruer. Washington, D.C.: Joseph Henry Press.

Anglin, J. M. 1977. *Word, Object, and Conceptual Development.* New York: Norton.

Belkin, M., and P. Niyogi. 2003. Laplacean Eigenmaps for dimensionality reduction and data representation. *Neural Computation* 15 (6):1373–1396.

Blanz, V., A. J. O'Toole, T. Vetter, and H. A. Wild. 2000. On the other side of the mean: The perception of dissimilarity in human faces. *Perception* 29 (8):885–891.

Briggman, K. L., and W. B. Kristan. 2006. Imaging dedicated and multifunctional neural circuits generating distinct behaviors. *Journal of Neuroscience* 26: 10925–10933.

Churchland, P. M. 1979. *Scientific Realism and the Plasticity of Mind.* Cambridge: Cambridge University Press.

Churchland, P. M., and C. A. Hooker. 1985. *Images of Science: Essays on Realism and Empiricism.* Chicago: University of Chicago Press.

Churchland, P. M. 1985. The ontological status of observables: In praise of the superempirical virtues. In P. M. Churchland and C. A. Hooker, *Images of Science.* Chicago: University of Chicago Press. Reprinted in P. M. Churchland, *A Neurocomputational Perspective* (Cambridge, MA: MIT Press, 1989).

Churchland, P. M. 1988. Perceptual plasticity and theoretical neutrality: A reply to Jerry Fodor. *Philosophy of Science* 55 (2):167–187.

Churchland, P. M. 1998. Conceptual similarity across sensory and neural diversity: The Fodor/Lepore challenge answered. *Journal of Philosophy* 95 (1): 5–32.

Churchland, P. M. 2001. Neurosemantics: On the mapping of minds and the portrayal of worlds. In *The Emergence of Mind*, ed. K. E. White. Milan: Fondazione Carlo Elba. Reprinted as chapter 8 of P. M. Churchland, *Neurophilosophy at Work* (New York: Cambridge University Press, 2007).

Churchland, P. M. 2007a. *Neurophilosophy at Work*. New York: Cambridge University Press.

Churchland, P. M. 2007b. On the reality (and diversity) of objective colors: How color-qualia space is a map of reflectance-profile space. *Philosophy of Science* 74 (2):119–149. Reprinted as chapter 10 of P. M. Churchland, *Neurophilosophy at Work* (New York: Cambridge University Press, 2007); and in *Essays in Honor of Larry Hardin*, ed. M. Matthen and J. Cohen (Cambridge, MA: MIT Press, 2010).

Churchland, P. M., and C. A. Hooker, eds. 1985. *Images of Science: Essays on Realism and Empiricism*. Chicago: University of Chicago Press.

Churchland, P. S., and C. L. Suhler. 2009. Control: Conscious and otherwise. *Trends in Cognitive Sciences* 13 (8):341–347.

Clark, A. 1998. *Being There: Putting Mind, Body, and World Together Again*. Cambridge: MIT Press.

Clark, A. 2003. *Natural Born Cyborgs*. Oxford: Oxford University Press.

Cottrell, G. 1991. Extracting features from faces using compression networks: Face, identity, emotion, and gender recognition using holons. In *Connectionist Models: Proceedings of the 1990 Summer School*, ed. D. Touretsky et al. San Mateo: Morgan Kaufmann.

Cottrell, G., and A. Laakso. 2000. Qualia and cluster analysis: Assessing representational similarity between neural systems. *Philosophical Psychology* 13 (1):77–95.

Cummins, R. 1997. The lot of the causal theory of mental content. *Journal of Philosophy* 94 (10):535–542.

Dawkins, R. 1976. *The Selfish Gene*. Oxford: Oxford University Press.

Dawkins, R. 1982. *The Extended Phenotype*. Oxford: Oxford University Press.

Dennett, D. C. 2003. *Freedom Evolves*. New York: Viking Books.

Fodor, J. A. 1975. *The Language of Thought*. New York: Thomas Y. Crowell.

Fodor, J. A. 1983. *The Modularity of Mind*. Cambridge, MA: MIT Press.

Fodor, J. A. 1990. *A Theory of Content and Other Essays*. Cambridge, MA: MIT Press.

Fodor, J. A. 2000. *The Mind Doesn't Work That Way: The Scope and Limits of Computational Psychology*. Cambridge, MA: MIT Press.

Fodor, J. A., M. Garrett, A. Garrett, F. Merrill, E. Walker, C. T. Parkes, and H. Cornelia. 1985. Against definitions. In *Cognition 8* Amsterdam: Elsevier Science. Reprinted in *Concepts: Core Readings*, ed. E. Margolis and S. Laurence, 491–512 (Cambridge, MA: MIT Press, 1999).

Fodor, J. A., and E. Lepore. 1992. Paul Churchland and state-space semantics. Chapter 7 of J. A. Fodor and E. Lepore, *Holism: A Shopper's Guide*. Oxford: Blackwell.

Fodor, J. A., and E. Lepore. 1999. All at sea in semantic space: Churchland on meaning similarity. *Journal of Philosophy* 8:381–403.

Garzon, F. C. 2000. State-space semantics and conceptual similarity: A reply to Churchland. *Philosophical Psychology* 13 (1):77–96.

Gettier, E. 1963. Is justified true belief knowledge? *Analysis* 23:121–123.

Giere, R. 2006. *Scientific Perspectivism*. Chicago: University of Chicago Press.

Goldman, A. 1986. *Epistemology and Cognition*. Cambridge, MA: Harvard University Press.

Graziano, M., C. S. Taylor, and T. Moore. 2002. Complex movements evoked by microstimulation of precentral cortex. *Neuron* 34: 841–851.

Grush, R. 1997. The architecture of representation. *Philosophical Psychology* 10 (1):5–25.

Hardin, L. 1993. *Color for Philosophers: Unweaving the Rainbow*. Indianapolis: Hackett.

Hebb, D. O. 1949. *The Organization of Behavior*. New York: Wiley.

Hooker, C. A. 1995. *Reason, Regulation, and Realism: Toward a Regulatory Systems Theory of Reason and Evolutionary Epistemology*. Albany, NY: SUNY Press.

Hopfield, J. J. 1982. Neural networks and physical systems with emergent collective computational abilities. *Proceedings of the National Academy of Sciences* 79: 2554–2558.

Hurvich, L. M. 1981. *Color Vision*. Sunderland, MA: Sinauer.

Hutchins, E. 1995. *Cognition in the Wild*. Cambridge, MA: MIT Press.

Jameson, D., and L. M. Hurvich, eds. 1972. *Visual Psychophysics*, vol. VII, no. 4, of *Handbook of Sensory Physiology*. Berlin: Springer-Verlag.

Johansson, G. 1973. Visual motion perception. *Scientific American* 232 (6):76–88.

Kristan, W. B., and K. L. Briggman. 2006. Imaging dedicated and multifunctional neural circuits generating distinct behaviors. *Journal of Neuroscience* 26: 10925–10933.

Kuhn, T. S. 1962. *The Structure of Scientific Revolutions*. Chicago: University of Chicago Press.

Lakatos, I. 1970. Falsification and the methodology of scientific research programmes. In I. Lakatos and A. Musgrave, *Criticism and the Growth of Knowledge*. Cambridge: Cambridge University Press.

Leopold, D. A., A. J. O'Toole, T. Vetter, and V. Blanz. 2001. Prototype-referenced shape encoding revealed by high-level aftereffects. *Nature Neuroscience* 4:89–94.

Mates, B. 1961. *Stoic Logic*. Berkeley: University of California Press.

O'Brien, G., and J. Opie. 2004. Notes towards a structuralist theory of mental representation. In *Representation in Mind*, ed. H. Clapin, P. Staines, and P. Slezak. Amsterdam: Elsevier.

O'Brien, G., and J. Opie. 2010. Representation in analog computation. In *Knowledge and Representation*, ed. A. Newen, A. Bartels, and E. Jung. Stanford, CA: CSLI Publications.

O'Toole, A. J., J. Peterson, and K. A. Deffenbacher. 1996. An "other-race effect" for categorizing faces by sex. *Perception* 25 (6):669–676.

Popper, K. 1972. Science: Conjectures and refutations. In K. Popper, *Conjectures and Refutations: The Growth of Scientific Knowledge*, 41–42. New York: Harper & Row.

Port, R. F., and T. van Gelder. 1995. *Mind as Motion: Explorations in the Dynamics of Cognition*. Cambridge, MA: MIT Press.

Quine, W. V. O. 1951. Two dogmas of empiricism. *Philosophical Review* 60: 20–43. Reprinted in W. V. O. Quine, *From a Logical Point of View* (Harvard University Press, 1953).

Rizzolatti, G., L. Fogassi, and V. Gallese 2001. Neurophysiological mechanisms underlying the understanding and imitation of action. *Nature Reviews: Neuroscience* 2 (9):661–670.

Rorty, R. 1979. *Philosophy and The Mirror of Nature*. Princeton, NJ: Princeton University Press.

Roweis, S. T., and S. K. Saul. 2000. Nonlinear dimensionality reduction by locally linear embedding. *Science* 290 (5500):2323–2326.

Rumelhart, D. E., and J. L. McClelland. 1986. On learning the past tenses of English verbs. In *Parallel Distributed Processing*, vol. 2, 216–271. Cambridge, MA: MIT Press.

Sejnowski, T. J., and S. Lehky. 1980. Computing shape from shading with a neural network model. In *Computational Neuroscience*, ed. E. Schwarz. Cambridge, MA: MIT Press.

Shepherd, R. N. 1980. Multidimensional scaling, tree-fitting, and clustering. *Science* 210:390–397.

Sherman, S. M. 2005. Thalamic relays and cortical functioning. *Progress in Brain Research* 149:107–126.

Sneed, J. D. 1971. *The Logical Structure of Mathematical Physics*. Dordrecht: Reidel.

Stegmuller, W. 1976. *The Structuralist View of Theories*. New York: Springer.

Suhler, C. L., and P. S. Churchland. 2009. Control: Conscious and otherwise. *Trends in Cognitive Sciences* 13 (8):341–347.

Tenenbaum, J. B., V. de Silva, and J. C. Langford. 2000. A global geometric framework for nonlinear dimensionality reduction. *Science* 290 (5500): 2319–2323.

Tiffany, E. 1999. Comments and criticisms: Semantics San Diego style. *Journal of Philosophy* 96 (8):416–429.

Usui, S., S. Nakauchi, and M. Nakano. 1992. Reconstruction of Munsell color space by a five-layer neural network. *Journal of the Optical Society of America* 9 (4):516–520.

van Fraassen, B. 1980. *The Scientific Image*. Oxford: Oxford University Press.

Van Gelder, T. 1993. Pumping intuitions with Watt's engine. *CogSci News* 6 (1):4–7.

Wang, J. W., A. M. Wong, J. Flores, L. B. Vosshall, and R. Axel. 2003. Two-photon calcium imaging reveals an odor-evoked map of activity in the fly brain. *Cell* 112 (2):271–282.

Zeki, S. 1980. The representation of colours in the cerebral cortex. *Nature* 284:412–418.

Index

Akins, K., 58, 281
Ampliative inference, 66–69, 140–141, 187–195, 232
Anglin, J., 84, 281
Anomalies, 222–223
Aristarchus, 186
Aristotle, 216
Armstrong, D., 153, 155
Autonomy, 151–153
Axel, R., 285

Bartels, A., 284
Bates, E., 23
Bechtel, B., 23
Belief, 30–31
Belkin, M., 144
Berkeley, G., 11, 78–79
Bernoulli, D., 22
Blanz, V., 284
Bohr, N., 206–207
Boltzmann, L., 21, 206, 228
Briggman, K. L., 281

Cameras, vii, 179–180, 245–248
Carnot, S., 267
Cartwright, N., 23, 24
Causal processes, 147–151, 153–157, 177–179, 186
Churchland, M. M., 172
Churchland, P. M., 54, 61, 93, 103, 111, 137, 186, 215, 251

Churchland, P. S., 152, 282
Clapin, H., 284
Clark, A., 274, 282
Cohen, J., 281
Color afterimages, 9
Color-constancy, 54–57
Color-space, 6–9, 50–61, 137
Conceptual redeployment, 20–24, 33–34, 273–274
Constructive empiricism, 215
Copernicus, N., 186, 194, 232, 235–236, 239
Cottrell, G., 7–8, 62–67, 72–74, 87–88, 112, 116, 139, 142, 164, 185, 187
Cronus, Diodorus, 155
Cummins, R., 97, 282

Darwin, C., 21, 189–192, 195, 272
Dawkins, R., 25, 264–265, 282
Deffenbacher, K. A., 136, 284
Dennett, D. C., 152, 282
Descartes, R., 5, 197
De Silva, V., 139, 144, 285
Dretske, F., 90
Dynamical redeployment of existing concepts, 21–25, 187–250

Einstein, A., 238–242
Elman, J., 23

Face gender-illusion, 8
Face-space, 7–8, 62–73
Fauconnier, G., 23
Flores, J., 285
Fodor, J. A., 5, 26, 28, 68–71, 84, 87, 90, 96–97, 115, 282–283
Forgassi, L., 284
Frame problem, 68–69
Freedom. *See* Autonomy

Galileo, 186
Gallese, V., 284
Garrett, M., 282
Garzon, F. C., 115
Gettier, E., 32, 283
Giere, R., 23, 24, 249
Goldman, A., 31, 105, 283
Graziano, M., 131, 283
Grush, R., 104, 283

Hardin, L., 137, 283
Hebb, D. O., 157–158
Hegel, G., 25, 261–262, 266–267, 277
Hesse, M., 23
Hooker, C. A., 255, 281, 282
Hopfield, J. J., 172, 283
Hume, D., 78, 83, 87–88, 90, 153, 197
Hurvich, L. M., 51–6, 61
Hutchins, E., 274, 283
Hypothetico-deduction, 197–200

Information compression, 61, 63, 142, 144–147
Intertheoretic reduction
  demeaning accounts of, 212–214
  earlier accounts of, 208–210
  the map-subsumption account of, 210–212

James, W., 83
Jameson, D., 51–56, 61, 283
Johansson, G., 140, 283

Johnson, M., 23
Jung, E., 284
Justification, 30–33

Kant, I., 1–5, 19, 79, 102–103, 126–129, 137–138, 184
Kepler, J., 194, 236, 236–237, 279
Kittay, E., 23
Kristin, W. B., 120, 281, 283
Kuhn, T. S., 23, 24, 195, 201, 222–223, 270–271

Laakso, A., 87, 116, 282
Lakatos, I., 200–201, 242, 270, 283
Lakoff, G., 23
Langacker, R., 23
Langford, J. C., 139, 144, 285
Language, 251–255, 260–261
Learning, 12–16
  supervised (back-propagation), 38–41, 143–144
  unsupervised (Hebbian), 157–165, 224–225, 246
  of temporal structures, 165
Lehky, S., 57, 284
Leibniz, G., 197
Leopold, D. A., 284
Lepore, E., 87
Locke, J., 11, 78–81, 81–84, 90, 137
Logical Empiricism, 261–264, 267, 272

Map-indexing theory of perception, ix, 98–102, 121, 127, 187, 221–222, 225–230, 250
Maps, vii–ix, 75–77, 81–82, 91–94, 99–103, 104–119, 123–138, 201–202, 205, 216–217, 236–238, 269–271
Mates, B., 155, 284
Matthen, M., 281
Maxwell, J. C., 22, 205–206, 242
McClelland, J. L., 71, 284
Meaning, identity of, 104–110
  similarity measure for, 110–113

Millikan, R., 90
Modal knowledge, 18
Moore, T., 131, 283
Motor control, 3–4, 136–137, 141–143, 150
Munsell, A. H., 50–51
Musgrave, A., 283

Nakano, M., 58–59, 285
Nakauchi, S., 58–59, 285
Nersessian, N., 23–24
Newen, A., 284
Newton, I., 21, 192–195, 197, 206, 213, 232, 236, 238
Normative vs. descriptive epistemology, 203–204
Nyogi, P., 144, 281

O'Brien, G., 76, 284
Opie, J., 76, 284
O'Toole, A. J., 10, 136, 281, 284

Peirce, C. S., 128–129, 132
Peterson, J., 136, 284
Plato, vii, ix, 153, 247
Popper, K., 197–200, 272
Port, R. F., 129, 132, 284
Preferred stimulus, 63–66, 86–87, 161
Ptolemy, 235, 239, 241

Quine, W. V. O., viii, 128, 155

Recurrent networks, 147–151, 165–179, 181–182, 191–192
Representation, viii, 4–6, 74–122
Rizzolatti, G., 49, 284
Rorty, R., 128–129, 132
Roweis, S. T., 144, 284
Rumelhart, D. E., 71, 284

Saul, K., 144, 284
Scientific realism, 134, 215
Sejnowski, T. J., 57, 284

Sellars, W., 128
Sheperd, R. N., 76, 284
Sherman, S. M., 175, 177, 284
Slezak, D., 284
Sneed, J. D., 24, 284
Staines, P., 284
Stegmuller, W., 24, 285
Suhler, C. L., 152, 282, 285

Taylor, C. S., 131, 283
Tenenbaum, J. B., 139, 144, 285
Theories
  sculpted activation-space view of, 22–24
  semantic view of, 22–24
  syntactic view of, 22–23
Tiffany, E., 108, 285
Torricelli, E., 21
Touretsky, D., 282
Truth, 30, 32
  correspondence theory of, 132–136

Usiu, S., 58–62, 142, 185, 285

van Frassen, B., 23, 24, 249, 285
Van Gelder, T., 129–132, 284
Vector completion, 66–69, 187–188
Vetter, T., 281, 284
Vosshall, L. B., 285

Wang, J. W., 119, 285
White, K. E., 281
Wild, H. A., 281
Wong, A. M., 285

Zeki, S., 83, 285